Collins

AQA GCSE
Maths

Higher Skills Book
Reason, interpret and communicate
mathematically and solve problems

Sandra Wharton

Contents

How to use this book

Welcome to the *AQA GCSE Maths Higher Skills Book*.

Focused on the new assessment objectives AO2 and AO3, this book is full of expertly written questions to build your skills in mathematical reasoning and problem solving.

This book is ideal to be used alongside the Practice Book or Student Book. It is structured by strand to encourage you to tackle questions without already knowing the mathematical context in which they sit – this will help prepare you for your exams.

Hints and tips

Some questions have a hint at the end of the chapter to get you started, but you should try to answer the question first, before looking at the hint. Take your time with the longer and multi-step questions – they are designed to make you think!

Worked exemplars

These give you suggested ways of working through these sorts of questions, step by step.

Answers

The answers are available online at www.collins.co.uk/gcsemaths4eanswers.

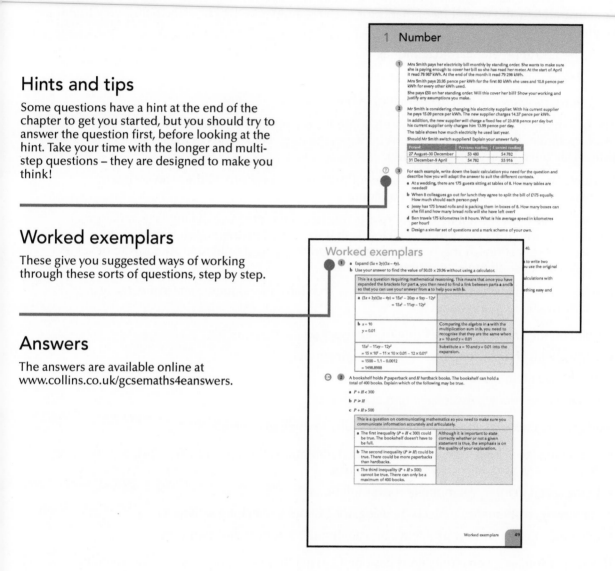

Teacher guide

- Build your students' confidence by tackling longer questions in class first, to generate discussions
- Questions are differentiated using a colour coded system from more accessible ⬤, through intermediate ⬤, to more challenging ⬤.
- Access the full mark scheme online

In the 2015 GCSE Maths specification there is an increased emphasis on problem solving, as demonstrated by the weighting of marks for each Assessment Objective (see below weightings for Higher):

AO1: Use and apply standard techniques	40%
AO2: Reason, interpret and communicate mathematically	30%
AO3: Solve problems within mathematics in other contexts	30%

The questions in this Skills Book draw on content from each of the strands to help students become more confident in making connections between the different strands of mathematics that have previously often been seen in isolation.

Students should also be encouraged to identify types of questions so they can access the mathematics they need quickly and confidently.

There is an increased emphasis in the revised curriculum on the use of language and correct terminology with the expectation at all levels that the use of mathematical language should be more precise – the questions in this book have addressed this, and should help students to develop a strong ability to communicate mathematically.

The approaches used in this book build on research into how students learn effectively in mathematics by requiring them to:

- build on the knowledge they bring to each section
- engage with and challenge common misconceptions
- emphasise methods rather than answers
- create connections between mathematical topics.

Students are encouraged to develop their ability to think mathematically by:

- evaluating the validity of statements and generalisations through activities such as classifying statements as sometimes, always or never true
- being asked to interpret multiple representations
- classifying mathematical objects by using, for example, matching activities
- creating and solving new problems, and analysing reasoning and solutions.

To support students, some questions have hints to get them started. These are indicated by ⑦. Build resilience by encouraging students to think carefully about the question before referring to these.

The Worked exemplars break down question types and show possible ways of tackling them in a step-by-step manner.

This book includes some longer and multi-step questions which start in a more straightforward way and get more challenging. Encourage students to take their time and think carefully about the information given to them, the answers required and the mathematics they will need to use.

Matching grid

Higher student book chapter and section	Question	Curriculum reference
1 Number: Basic number		
1.1 Solving real-life problems	p10 Q1 , Q2 , Q3 , p11 Q10	N2
1.2 Multiplication and division with decimals	p10 Q4 , p13 Q19, Q20	N10
1.3 Approximation of calculations	p14 Q23–Q26	N15
1.4 Multiples, factors, prime numbers, powers and roots	p11 Q5 , Q6	N4
2 Number: Fractions, ratio and proportion		
2.2 Adding, subtracting and calculating with fractions	p13 Q16	N1, N2, N8, N10
2.3 Multiplying and dividing fractions	p13 Q17, p44 Q11, p46 Q16	N2, N10
2.4 Fractions on a calculator		N10
2.5 Increasing and decreasing quantities by a percentage	p43 Q3 , p45 Q13–Q15, p46 Q16, p59 Q5	N3, N12, R9
3 Statistics: Statistical diagrams and averages		
3.1 Statistical representation	p89 Q7 , p95 Q31	S2
3.2 Statistical measures	p88 Q5 , Q6 , p89 Q10, Q11, p90 Q12, Q14, Q16, Q17, p91 Q18, Q19	S4
3.3 Scatter diagrams	p96 Q32, Q33	S6
4 Algebra: Number and sequences		
4.2 Number sequences	p33 Q62	A23
4.3 Finding the nth term of a linear sequence	p30 Q43, Q44, p31 Q45, p33 Q58	A25
4.4 Special sequences	p30 Q41, p33 Q53, Q59–Q61	A24
4.5 General rules from given patterns	p29 Q40, p30 Q41, Q42	A25
4.6 The nth term of a quadratic sequence	p29 Q39	A25
4.7 Finding the nth term for quadratic sequences	p29 Q39, p30 Q42–Q44, p31 Q46, p32 Q51	A25
5 Ratio, proportion and rates of change: Ratio and proportion		
5.1 Ratio	p43 Q1 , p44 Q7 , Q11, p60 Q14	R4
5.2 Direct proportion problems	p45 Q12, p46 Q15, Q17–Q19, p47 Q20, Q22–Q24, p48 Q25	R10
5.3 Best buys	p11 Q9 , p43 Q2 , p44 Q8 –Q10, p45 Q13	R10
5.4 Compound measures	p43 Q4 –Q6 , Q5 , p47 Q21	R11
5.5 Compound interest and repeated percentage change	p48 Q28, Q29, p51 Q37, Q38, p52 Q39	R16
5.6 Reverse percentage (working out the original amount)	p45 Q14, p46 Q16, Q17, p48 Q26	R9

Higher student book chapter and section	Question	Curriculum reference
6 Geometry and measures: Angles		
6.2 Triangles	p59 Q1 , Q3 , p60 Q14, p61 Q17, p66 Q57, p73 Q84	G4, G6
6.3 Angles in a polygon	p60 Q10 , Q13	G3
6.4 Regular polygons	p60 Q10 , Q11 , p61 Q16, p65 Q52, p66 Q58	G3
6.5 Angles in parallel lines	p65 Q49, p66 Q58	G3
6.6 Special quadrilaterals	p59 Q2 , p60 Q12, p61 Q18, p65 Q51, p66 Q56	G4
6.7 Scale drawings and bearings	p62 Q30, p64 Q44, p72 Q79	G15
7 Geometry and measures: Transformations, constructions and loci		
7.1 Congruent triangles	p60 Q7 , Q8 , p65 Q49, Q50	G5
7.2 Rotational symmetry	p25 Q21, p68 Q61	G7
7.3 Transformations	p59 Q4 , p61 Q20–Q22, p67 Q59, Q60	G7
7.4 Combinations of transformations	p67 Q59, p68 Q61, p73 Q85	G8
7.5 Bisectors	p64 Q46, p65 Q47, Q48	G2
7.6 Defining a locus	p60 Q9	G2
7.7 Loci problems	p60 Q9	G2
7.8 Plans and elevations	p62 Q26	G13
8 Algebra: Algebraic manipulation		
8.1 Basic algebra	p22 Q1 – Q6 , p24 Q12, Q13, p25 Q17, p29 Q38, p33 Q54–Q56, Q62, Q63, p34 Q66, Q72, p73 Q84	A1
8.2 Factorisation	p25 Q16	A4
8.3 Quadratic expansion	p31 Q47, p33 Q52, Q57, Q64, Q65, p34 Q66, Q71, p35 Q74	A4
8.4 Expanding squares	p34 Q68	A4
8.6 Quadratic factorisation	p34 Q67	A4
8.7 Factorising $ax^2 + bx + c$	p34 Q67	A4
8.8 Changing the subject of a formula	p24 Q14	A5
9 Geometry and measures: Length, area and volume		
9.1 Circumference and area of a circle	p60 Q15, p61 Q25, p62 Q27, p66 Q57	G17
9.2 Area of a parallelogram	p72 Q82	G16
9.4 Sectors	p61 Q24	G18
9.5 Volume of a prism	p59 Q5 , p61 Q23, p62 Q31, p69 Q69, p70 Q70, Q71, Q72, p71 Q73, Q76	G16
9.6 Cylinders	p60 Q6 , p71 Q73	G16

Higher student book chapter and section	Question	Curriculum reference
9.7 Volume of a pyramid	p70 Q71, p72 Q77	G17
9.9 Spheres	p70 Q71	G17
10 Algebra: Linear graphs		
10.2 Gradient of a line	p25 Q19	A10
10.3 Drawing graphs by gradient-intercept and cover-up methods	p25 Q18	A9
10.4 Finding the equation of a line from its graph	p49 Q30	A9
10.5 Real-life uses of graphs	p23 Q8 –Q10, p24 Q11, p25 Q19, Q22	A14
10.7 Parallel and perpendicular lines	p27 Q28	A9
11 Geometry: Right-angled triangles		
11.1 Pythagoras' theorem	p29 Q38, p33 Q52, p61 Q19, p62 Q28, p63 Q37, p66 Q54, Q55, p69 Q66, p73 Q83, Q84	G20
11.2 Finding the length of a shorter side	p64 Q38	G20
11.3 Applying Pythagoras' theorem in real-life situations	p63 Q34, Q35, p72 Q80, Q82	G20
11.4 Pythagoras' theorem and isosceles triangles	p63 Q34	G20
11.5 Pythagoras' theorem in three dimensions	p72 Q78	G20
11.8 Using the sine and cosine functions	p63 Q36, Q37, p64 Q41, Q42, p72 Q80	G20
11.9 Using the tangent function	p64 Q41	G20
11.10 Which ratio to use	p61 Q19	G20
11.11 Solving problems using trigonometry	p62 Q29, p64 Q41, Q42, Q43	G20
11.12 Trigonometry and bearings	p72 Q79	G20
12 Geometry and measures: Similarity		
12.2 Areas and volumes of similar shapes	p47 Q22–Q24, p48 Q25, Q26, p63 Q33, p65 Q51, p66 Q56, Q57	G19
13 Probability: Exploring and applying probability		
13.1 Experimental probability	p80 Q1, Q2	P1, P2, P3, P5
13.2 Mutually exclusive and exhaustive outcomes	p80 Q4, p81 Q5	P4
13.3 Expectation	p80 Q3	P5
13.4 Probability and two-way tables	p81 Q7, Q8	P6
13.5 Probability and Venn diagrams	p83 Q15, Q16	P6
14 Number: Powers and standard form		
14.1 Powers (indices)	p12 Q13, p13 Q18, p15 Q33	N6, N7
14.2 Rules for multiplying and dividing powers	p15 Q34	N6, N7
14.3 Standard form	p12 Q12, Q15, p15 Q28, Q32, p16 Q38	N9

Higher student book chapter and section	Question	Curriculum reference
15 Algebra: Equations and inequalities		
15.1 Linear equations	p22 Q3, p24 Q14, Q15	A17
15.2 Elimination method for simultaneous equations	p27 Q29–Q32, p28 Q33	A19
15.3 Substitution method for simultaneous equations	p27 Q29–Q32, p28 Q33	A19
15.4 Balancing coefficients to solve simultaneous equations	p27 Q29–Q32, p28 Q33	A19
15.5 Using simultaneous equations to solve problems	p27 Q32, p28 Q33, Q35, Q36	A19
15.6 Linear inequalities	p32 Q48	A22
15.7 Graphical inequalities	p32 Q49, Q50	A22
16 Number: Counting, accuracy, powers and surds		
16.1 Rational numbers, reciprocals, terminating and recurring decimals	p16 Q39, Q40	N10
16.2 Estimating powers and roots	p15 Q29	N7, N14
16.3 Negative and fractional powers	p12 Q14, p15 Q28, Q30, Q31, p33 Q63	N7
16.4 Surds	p15 Q28, Q35, Q36, p16 Q37, p64 Q38	N8
16.5 Limits of accuracy	p13 Q21, p14 Q22–Q26, p16 Q41	N16
16.6 Problems involving limits of accuracy	p11 Q7, Q8, p63 Q36	N16
16.7 Choices and outcomes	p11 Q11, p15 Q27, p22 Q7	N5
17 Algebra: Quadratic equations		
17.1 Plotting quadratic graphs	p25 Q20, Q21, p27 Q26, p36 Q76, p49 Q30	A12
17.2 Solving quadratic equations by factorisation	p28 Q37	A18
17.4 Solving quadratic equations by completing the square	p36 Q77	A18
17.5 The significant points of a quadratic curve	p27 Q25, Q26, p35 Q75	A11
17.6 Solving one linear and one non-linear equation using graphs	p28 Q34	A19
18 Statistics: Sampling and more complex diagrams		
18.1 Collecting data	p88 Q1, Q2 – Q4, p90 Q15	S1
18.2 Frequency polygons	p89 Q8	S3
18.3 Cumulative frequency graphs	p90 Q13, p91 Q20, p92 Q21, Q23, p93 Q27	S3
18.4 Box plots	p92 Q24, p93 Q25–Q27	S4
18.5 Histograms	p89 Q9, p92 Q22, p94 Q28, Q29, p95 Q30	S3
19 Probability: Combined events		
19.1 Addition rules for outcomes of events	p82 Q11	P8
19.2 Combined events	p81 Q7, Q8, p82 Q12, p83 Q17	P7, P8
19.3 Tree diagrams	p81 Q6, p82 Q9, Q10, Q13, p83 Q20	P6, P8

Higher student book chapter and section	Question	Curriculum reference
19.4 Independent events	p82 Q14, p83 Q18, p84 Q24–Q26	P8
19.5 Conditional probability	p83 Q19, Q21, Q22, p84 Q23	P9
20 Geometry and measures: Properties of circles		
20.1 Circle theorems	p68 Q64, p69 Q65, Q68	G10
20.2 Cyclic quadrilaterals	p69 Q65, Q67	G10
20.3 Tangents and chords	p68 Q63, p69 Q66, Q67	G10
21 Ratio, proportion and rates of change: Variation		
21.1 Direct proportion	p48 Q27, p50 Q33, p69 Q69	R13
21.2 Inverse proportion	p50 Q31, Q32	R13
22 Geometry and measures: Triangles		
22.1 Further 2D problems	p72 Q82, p73 Q83	G20
22.2 Further 3D problems	p71 Q76, p72 Q77, Q78	G20
22.3 Trigonometric ratios of angles between 0° and 360°	p64 Q39, Q40	G21
22.4 Solving any triangle	p16 Q41, p61 Q19, p64 Q43, p66 Q53, p71 Q74, Q75, p72 Q81, Q82, p73 Q83	G22
22.5 Using sine to calculate the area of a triangle	p34 Q72, p72 Q81, Q82, p73 Q83	G23
23 Algebra: Graphs		
23.1 Distance–time graphs	p23, Q8	A15
23.4 Rates of change	p50 Q34, Q35, p51 Q36	A15
23.5 Equation of a circle	p27 Q28	A16
23.6 Other graphs	p25 Q20, Q21, p26 Q23, p28 Q37, p49 Q30, p52 Q40–Q43	A12, A14
23.7 Transformations of the graph $y = f(x)$	p26 Q24, p36 Q76	A13
24 Algebra: Algebraic fractions and functions		
24.1 Algebraic fractions	p25 Q16, p34 Q69, Q70	A4
24.2 Changing the subject of a formula	p24 Q14, p53 Q44	A5
24.3 Functions	p34 Q73, p36 Q76, p50 Q34, Q35, p51 Q36	A7
24.5 Iteration	p53 Q44	A20
25 Geometry and measures: Vector geometry		
25.1 Properties of vectors	p64 Q45, p67 Q59	G25
25.2 Vectors in geometry	p64 Q44, p73 Q86, Q87, p74 Q88–Q90	G24, G25

basic

medium

hard

1 Number

1. Mrs Smith pays her electricity bill monthly by standing order. She wants to make sure she is paying enough to cover her bill so she has read her meter. At the start of April it read 78 987 kWh. At the end of the month it read 79 298 kWh.

   Mrs Smith pays 20.95 pence per kWh for the first 80 kWh she uses and 10.8 pence per kWh for every other kWh used.

   She pays £50 on her standing order. Will this cover her bill? Show your working and justify any assumptions you make.

2. Mr Smith is considering changing his electricity supplier. With his current supplier he pays 15.09 pence per kWh. The new supplier charges 14.37 pence per kWh.

   In addition, the new supplier will charge a fixed fee of 23.818 pence per day but his current supplier only charges him 13.99 pence per day.

   The table shows how much electricity he used last year.

   Should Mr Smith switch suppliers? Explain your answer fully.

Period	Previous reading	Current reading
27 August–30 December	53 480	54 782
31 December–9 April	54 782	55 916

3. For each example, write down the basic calculation you need for the question and describe how you will adapt the answer to suit the different contexts.

   a At a wedding, there are 175 guests sitting at tables of 8. How many tables are needed?

   b When 8 colleagues go out for lunch they agree to split the bill of £175 equally. How much should each person pay?

   c Jessy has 175 bread rolls and is packing them in boxes of 8. How many boxes can she fill and how many bread rolls will she have left over?

   d Ben travels 175 kilometres in 8 hours. What is his average speed in kilometres per hour?

   e Design a similar set of questions and a mark scheme of your own.

4. a For each question, explain what you are doing.

   i Use the facts $4.6 \times 10 = 46$ and $46 \times 4 = 184$ to work out 4.6×40.

   ii Use estimation to check that your answer is reasonable.

   iii You know that $115.6 \div 3.4 = 34$. Explain how you can use this to write two different calculations with the same answer. Explain how you use the original calculation to get to the new calculations.

   iv $24 \times 72 = 1728$. Explain how you can use this fact to devise calculations with answers 17.28.

   b Design your own question like those in part a. Start with something easy and progress to something more difficult.

5 Give reasons why none of these numbers are prime numbers.

2484, 17 625, 3426

6 629 is a product of two prime numbers. One of the numbers is 17.
What is the other one? Explain how you know.

7 A parking space is 4.8 metres long, to the nearest tenth of a metre.

A car is 4.5 metres long, to the nearest half metre.

Which of the following statements is definitely true?

A: The space is big enough.

B: The space is not big enough.

C: It is impossible to tell whether the space is big enough.

Explain how you decide.

8 Billy has 20 identical bricks. Eack brick is 15 cm long, to the nearest centimetre.

The bricks are put end to end to build a wall. What is the greatest possible length of the wall?

9 Barry has a four-wheel drive vehicle. It uses a lot of petrol. His garage says he can convert the car to a different type of fuel called LPG, which is much cheaper than petrol. However, it will cost him £2500 to convert the car so it can use this fuel.

a Write a question for Barry to answer to help him decide if he is going to convert his car.

b What assumptions or pieces of information will Barry need to make, to answer his question?

c Using your answer to parts **a** and **b**, present an argument to help Barry decide if he should get his car converted.

d Barry has to take out a loan to pay for the conversion. He pays £66.19 over 60 months. Does this change the answer to part **c**?

10 A dolphin has to breathe even when it is asleep in the water. Alison is a marine biologist who studies dolphins. She observes that a dolphin takes a breath at the surface, dives to the bottom of the sea and starts to sleep.

From the bottom it floats slowly to the surface in 5 minutes and then takes another breath.

After another 2 minutes it is back at the bottom of the sea.

Alison watches the dolphin for 1.5 hours. At the end of the 1.5 hours was the dolphin at the bottom, on its way up, breathing or on its way down?

Explain your answer.

11 **a** How many ways can you label a product when the label must consist of a letter AND a number from 1 to 25.

b A restaurant has five different flavours of ice cream, in four different size servings, and a choice of a cone or a tub. How many different possible ways are there of serving the ice cream?

12　a Copy the table and rewrite the distance and the diameter in standard form.

Planet	Distance from the Sun (million km)	Mass (kg)	Diameter (km)
Mercury	58	3.3×10^{23}	4878
Venus	108	4.87×10^{24}	12 104
Earth	150	5.98×10^{24}	12 756
Mars	228	6.42×10^{23}	6787
Jupiter	778	1.90×10^{27}	142 796
Saturn	1427	5.69×10^{26}	120 660
Uranus	2871	8.68×10^{25}	51 118
Neptune	4497	1.02×10^{26}	48 600
Pluto	5913	1.29×10^{22}	2274

b i Which planet is the largest?

　ii Which planet is the smallest?

　iii Which planet is the lightest?

　iv Which planet is the heaviest?

　v Which planet is approximately twice as far away from the Sun as Saturn is?

　vi Which two planets are similar in size?

c Can you identify a relationship between each planet's mass, distance from the Sun and diameter? Are there any trends? Write a short paragraph to describe your initial findings.

13　Are the following statements always, sometimes or never true? If a statement is sometimes true, when is the statement true and when is it false?

a Cubing a number always gives a result that is bigger than the number.

b The square of a number is always positive.

c You can find the square root of any number.

d You can find the cube root of any number.

14　a Show that:

　i $5^6 \div 5^{-3} = 5^9$

　ii $5^6 \times 5^{-3} = 5^3$

b What does the index of $\frac{1}{2}$ represent?

c What are the values of n and c in the equation $27 \times 48 = n^4 \times 2^c$?

15　The diameter of the smallest virus is 20 nanometres. The average area of the pupil of a human eye in light is $9\pi \times 10^{-6}$ m². In 2014, the tallest skyscraper was the Burj Khalifa in Dubai. It is 8.298×10^2 m high. The height of Mount Everest is 8.848×10^3 m.

a How many times taller is the mountain than the skyscraper?

b How high is the skyscraper in kilometres (km)?

c A nanometre is $\frac{1}{1\,000\,000\,000}$ of a metre. Write the diameter of the virus in metres in standard form.

d What is the diameter of the virus as a fraction of the diameter of the human eye? Give your answer in standard form correct to 3 sf.

16 The diagram shows three identical shapes A, B and C.

$\frac{5}{7}$ of shape A is shaded.

$\frac{8}{9}$ of shape C is shaded.

What fraction of shape B is shaded?

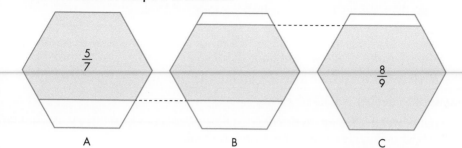

A   B   C

17 In a city $5\frac{1}{2}$ out of every 15 square metres are used for residential development and services. This includes services such as schools, doctors' surgeries and play areas. $\frac{5}{8}$ of the residential land is actually used for housing. What percentage of the total area of the city is used for these services?

18 Siobhan runs a company that manufactures small items out of plastic. One item the company makes is a solid plastic dice with a side of 2 cm.

Siobhan wants to make a larger dice. She needs to know what it will cost. This will depend on the amount of plastic required to make it.

a Siobhan thinks that a dice with a side of 4 cm will require twice as much plastic to make as a dice with a side of 2 cm. Explain why this is not the case. How much more plastic will be required?

b Siobhan is thinking about making a dice with a side of 3 cm.

    i How much more plastic will this use, compared to a dice with a side of 2 cm?

    ii How much less plastic will a dice with a side of 3 cm require, compared to a dice with a side of 4 cm?

c What advice can you give Siobhan about the size of a dice that requires twice as much plastic to make as one that has a side of 2 cm?

19 **a** Will the answer to $68 \div 0.8$ be larger or smaller than 68? Justify your answer.

b Estimate an answer to each calculation. Is your estimate higher or lower that the real answer? Explain your answer.

    i 75.8×23.1

    ii $\frac{26.2}{3.5}$

    iii $(1.9)^2 \times 7.4$

20 **a** Give three examples of multiplication and division calculations, with answers that approximate to 75. Try to make your examples progressively more complex. Explain how you decided on each of the calculations.

b Why is $9 \div 3$ a better approximation for $9.58 \div 3.46$ than $10 \div 3$?

21 The dimensions of a classroom floor, measured to the nearest metre, are 19 m by 15 m. What range must the area of the floor lie within? Suggest a sensible answer for the area, given the degree of accuracy of the measurements.

22 You are the manager of a haulage company.

You own a lorry with six axles, with a maximum axle weight limit of 10.5 tonnes.

You also own a lorry with five axles, with a maximum axle weight limit of 11.5 tonnes.

This lorry is 10% cheaper to run per trip than the six-axle lorry.

These are the load restrictions for heavy goods vehicles.

Type of lorry	Load limit
six axles, maximum axle weight limit 10.5 tonnes	44 tonnes
five or six axles, maximum axle weight limit 11.5 tonnes	40 tonnes

You want to deliver pallets of goods.

Each pallet weighs 500 kg, to the nearest 50 kg. Explain which lorry you would use for each job, showing clearly any decisions you make.

a A company orders 80 pallets.

b A company orders 150 pallets.

c A company orders 159 pallets.

23 **a** Explain the difference in meaning between 0.4 m and 0.400 m. Why is it sometimes necessary to include zeros in measurements?

b What range of measured lengths might be represented by the measurement 430 cm?

c What accuracy is needed to be sure a measurement is accurate to the nearest metre?

d Why might you pick a runner whose time for running 100 m is recorded as 13.3 seconds rather than 13.30 seconds? Why might you not?

e Explain how seven people each with a mass of 100 kg might exceed a limit of 700 kg for a lift.

24 A theatre has 365 seats.

For a show, 280 tickets are sold in advance.

The theatre's manager estimates that another 100 people, to the nearest 10, will turn up without tickets.

Is it possible they will all get a seat, assuming that 5% of those with tickets do not turn up?

Show clearly how you decide.

25 A stopwatch records the time for a winner of a 100 metre race as 12.3 seconds, measured to the nearest one-tenth of a second.

The length of the track is correct to the nearest centimetre.

What is the greatest possible average speed of the winner?

26 A cube has a volume of 125 cm³, to the nearest 1 cm³. Find the limits of accuracy of the area of one side of the square base.

27 a Are these statement true or false? Explain your answer.

    i There are $n > k$ balls and the balls are divided between k boxes. At least one box contains at least two balls.

    ii When there are $n > k$ balls, at least one box will have $\frac{n}{k}$ balls.

    iii In a group of 366 people no one has the same birthday.

b There are six possible grades in a test: A, B, C, D, E and F. How many students are needed to be sure at least five students get the same grade?

28 Which of these statements are true? Justify your answer in each case.

a the length of an A4 piece of paper is 2.97×10^5 km

b $3^{-3} = \frac{1}{3} - 9$

c $16^2 = 2^8$

d $4\sqrt{3} \times 3\sqrt{3} = 7\sqrt{7}$

29 Without using a calculator, decide whether these statements are true or false. Justify you answer for each statement.

a $\sqrt{19}$ is greater than 5

b $\sqrt{23}$ is between 4 and 5

c $2\sqrt{2}$ is less than $\sqrt{8}$

d $\sqrt{0.38}$ is greater than 0.6

30 Which of these is the odd one out?

$27^{-\frac{1}{3}}$ $25^{-\frac{1}{2}}$ 3^{-1}

Show how you decided.

31 $x^{-\frac{1}{4}} = y^{-\frac{1}{2}}$

Find values for x and y that make this equation work.

32 A number is between 10 000 and 100 000. It is written in the form 2.5×10^n.

What is the value of n?

33 Write down the value of each of these.

a $\sqrt{0.36}$ **b** $\sqrt{0.81}$ **c** $\sqrt{1.69}$ **d** $\sqrt{0.09}$ **e** $\sqrt{0.01}$

f $\sqrt{1.44}$ **g** $\sqrt{2.25}$ **h** $\sqrt{1.96}$ **i** $\sqrt{4.41}$ **j** $\sqrt{12.25}$

34 Decide whether this statement is true or false.

$\sqrt{a^2 + b^2} = a + b$

Show your working.

35 Write down a division of two different surds that has an integer answer.

Show your working.

36 a Write down two surds that, when added, give a rational number.

b Write down two surds that, when added, do not give a rational number.

37 Katie is working out the height of a curtain for a window, in metres. Her calculator displays the answer $1 + \sqrt{3}$.

Without using a calculator, explain why the height of the curtain is between 2 metres and 3 metres.

38 The Earth can be modelled as a sphere with diameter 1.2756×10^4 km. In August 2014 the population of the planet was $7.185\,004 \times 10^9$ to the nearest thousand.

a If everyone spread out on the surface of the Earth, what area of the planet would each person have?

b Approximately 70% of the surface area of the Earth is water. How does this affect your answer to part **a**?

39 Are these statements true or false? Justify your answers.

a All terminating decimals can be written as fractions.

b All recurring decimals can be written as fractions.

c Any number can be written as a fraction.

40 **a** Which of these fractions are equivalent to terminating decimals? Do not work them out. Explain your reasoning.

$$\frac{3}{5}, \frac{3}{11}, \frac{7}{30}, \frac{9}{22}, \frac{9}{20}, \frac{7}{16}$$

b $\frac{1}{6}$ is a recurring decimal. What other fractions related to one-sixth will also be recurring?

c Using the knowledge that $\frac{1}{6} = 0.166\,666...$, explain how you would find the decimal equivalents of $\frac{1}{3}$ and $\frac{1}{60}$.

d Which of these recurring decimals would be easy to convert? Are any of them difficult to convert? What makes them easy or difficult to convert?

$0.027\,272\,7...,\ 0.272\,727...,\ 2.727\,272...,\ 27.272\,727...$

e Can you use the fraction equivalents of 2.72... and 27.27.... to prove the second is ten times greater than the first?

41 In the triangle ABC, the length of side AB is 42 cm to the nearest centimetre.

The length of side AC is 35 cm to the nearest centimetre.

The angle C is 61° to the nearest degree.

What is the largest possible size that angle B could be?

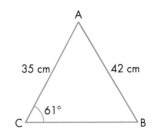

⑦Hints and tips

Question	Hint
3	Think about what a sensible answer is. Can you have part of a table or a fraction of a penny?
9	Think about issues such as how long it will take to recover the cost of converting the vehicle.
27	Start with a simple example and a diagram.
38	The surface area of a sphere is $4\pi r^2$.
41	You will need to use the sine rule.

Worked exemplars

PS **1** I earn £30 000 in 12 months.

20% of this is deducted for tax.

$x\%$ is deducted for National Insurance.

At the end of each month I have £1800 left.

Work out the value of x.

This is a problem-solving question. You will need to plan a strategy to solve it and, most importantly, communicate your method clearly.	
Method 1 Yearly take home pay = £1800 × 12 = £21 600 Amount deducted = £30 000 − £21 600 = £8400 Tax deducted = £30 000 × 0.2 = £6000 NI deducted = £8400 − £6000 = £2400 NI as a percentage of earnings = 2400 ÷ 30 000 × 100 = 0.08 × 100 $x = 8\%$	There are two different methods that lead to the correct answer: method 1 works on yearly earnings and method 2 on monthly earnings. Use the one you find easier to explain.
Method 2 £30 000 ÷ 12 = £2500 £2500 − £1800 = £700 £2500 × 0.2 = £500 £700 − £500 = £200 £200 ÷ £2500 × £100 = 8%	*Do not* just write down numbers. Get used to writing, in words, what you are calculating. Copy these calculations and put some words at the start of each line to explain what is being worked out.

MR **2** Two numbers have been rounded.

The first number is 360 to two significant figures.

The second number is 500 to one significant figure.

What is the smallest possible sum of the two original numbers?

This question assesses your mathematical reasoning.	
The smallest that the first number could be is 355. The smallest that the second number could be is 450.	First work out the smallest possible value for each number.
Smallest sum = 355 + 450 = 805	Then work out the sum of these two numbers. Sometimes you may make a mistake, say working out the smallest value of the second number as 495, but as long as you have explained your working you can still get some credit. That is why it is essential to write down what you are working out.

3 Two numbers, x and y, can be written in prime factor form as $x = 2^3 \times a \times b^2$ and $y = 2^2 \times a^3 \times b$, where a and b are prime numbers greater than 2.

a Which of these terms are factors of both x and y?

$2ab$ 4 $8a^2b$ $4a^2b^2$

b Which of these expressions is the LCM of x and y?

$2ab$ $8a^3b^2$ $4ab$ $32a^4b^3$

This is a mathematical reasoning question with a multi-choice answer, so there is no working to be shown.	
a $2ab$ and 4	Look at each expression and decide if it will divide into x and y. $2ab$ and 4 will divide into both but a^2 will not divide into x, and $8b^2$ will not divide into y.
b $8a^3b^2$	Take the number term and each letter term in turn. There must be at least 2^3, at least a^3 and at least b^2 in the LCM.

  4 **a** Work this out.

$$\frac{\left(\frac{2}{3} + \frac{4}{5}\right)}{1\frac{7}{9}}$$

b Decide whether $\frac{2}{3} + \frac{4}{5}$ is greater or less than $1\frac{7}{9}$.
Show clearly how you decide.

For part **a** you need to use basic mathematics skills. Then **b** is the 'communicating mathematics' part of the question. You need to make your method clear.	
a $\frac{2}{3} + \frac{4}{5} = \frac{10}{15} + \frac{12}{15}$ $= \frac{22}{15}$	First add the two fractions inside the brackets by writing them with a common denominator, that is, $\frac{10}{15} + \frac{12}{15}$.
$\frac{22}{15} \times \frac{9}{16}$	Change the mixed number $1\frac{7}{9}$ into an improper fraction, $\frac{16}{9}$, then find its reciprocal and multiply.
$\frac{^{11}22}{_5 15} \times \frac{9^3}{16_8} = \frac{33}{40}$	Cancel the common factors. Then multiply the numerators and multiply the denominators.
b The answer to part **a** $\left(\frac{33}{40}\right)$ is less than 1. This means that the numerator $\left(\frac{2}{3} + \frac{4}{5}\right)$ must be smaller than the denominator $\left(1\frac{7}{9}\right)$.	This is a communicating mathematics question so you need to make sure your use your answer to part **a** to decide whether $\frac{2}{3} + \frac{4}{5}$ is greater or less than $1\frac{7}{9}$ and give a reason to support your answer.

5 This is a table of powers of 3.

3^1	3^2	3^3	3^4	3^5	3^6	3^7
3	9	27	81	243	729	2187

a Use your calculator to work out $27 \div 243$. Give the answer as a fraction.

b Use the rules of indices to write $3^3 \div 3^5$ as a single power of 3.

c Deduce the value, as a fraction, of 3^{-3}.

This is a mathematical reasoning question. The first two parts set up the information you will need.	
a $\dfrac{27}{243} = \dfrac{1}{9}$	Write $27 \div 243$ as a fraction, then cancel to the simplest form. Make sure you know how to change an answer into a fraction if the display shows a decimal, in this case 0.111…
b $3^3 \div 3^5 = 3^{3-5} = 3^{-2}$	Apply the Ω rules of indices. When dividing powers with the same base, subtract them.
c **a** and **b** $\Rightarrow \dfrac{1}{9} = 3^{-2}$ $\therefore 3^{-3} = \dfrac{1}{27}$	This is where the mathematical reasoning comes in. Parts **a** and **b** are linked in that they are the same calculation in different forms, so the answers must be the same. So if $\frac{1}{9}$ = 3^{-2}, then 3^{-3} must be $\frac{1}{27}$. Remember that the symbol $\Rightarrow$ means 'implies' and Ω means 'therefore'.

6 The population of the world is approximately 7 billion.

One grain of sand has a mass of 0.0026 grams.

2.6 grams of sand have a total volume of 1 cm³.

Work out the size of a cube that would be big enough to hold as many grains of sand as the population of the world.

This problem-solving question requires you to translate a real-life problem into a series of mathematical processes.	
Number of grains of sand in $1 \text{ cm}^3 = 2.6 \div 0.0026 = 1000$ or 10^3	First work out how many grains of sand there are in 1 cm³. Don't forget to write down what you are working out.
Number of cubic centimetres that would hold 7 billion grains $= 7 \times 10^9 \div 10^3$ $= 7 \times 10^6$	Next work out how many cubic centimetres would hold 7 billion grains.
Side of cube $= \sqrt[3]{7 \times 10^6}$ $= 191 \text{ cm}$ $\approx 2 \text{ m}$	Now work out the side of the cube by finding the cube root of the answer. You can leave the answer in centimetres or convert to metres. The answer is surprisingly small. Remember that $\approx$ means 'approximately'. Rounding the answer is acceptable, as long as you show working, as all values are approximations.

7 Four-digit numbers are to be made using four of these five number cards.

| 1 | 2 | 3 | 4 | 5 |

Show clearly that the number of even four-digit numbers between 3000 and 5000 is 18.

> This is a 'communicating mathematically' question so make it clear, using words, what you are doing.

The even four-digit numbers between 3000 and 5000 will be 3■■2 or 3■■4 or 4■■2, where ■■ are any two of the remaining three digits. The number of ways of picking 2 digits from 3 where the order matters is $3 \times 2 = 6$. As there are 3 possible sets and each set has 6 possible ways, this is $3 \times 6 = 18$.	Show that you understand where the numbers come from. Explain the number of ways of getting 2 digits from 3. Even though it is obvious that $3 \times 6 = 18$, explain the final step.

  8 The area of this rectangle is $(12 - 3\sqrt{2})$ cm².

Work out the perimeter of the rectangle. Give your answer in the form $a\sqrt{2} \pm b$, where a and b are integers.

$3\sqrt{2}$ cm

> This is a problem-solving question so you will need to show your strategy.
>
> You need to establish the missing side first. There are two possible methods.

Method 1 $\dfrac{12 - 3\sqrt{2}}{3\sqrt{2}} = \dfrac{(12 - 3\sqrt{2}) \times 3\sqrt{2}}{3\sqrt{2} \times 3\sqrt{2}}$	Divide the area by $3\sqrt{2}$, then rationalise the denominator.
$= \dfrac{36\sqrt{2} - 9 \times 2}{9 \times 2}$	Simplify, then factorise.
$= \dfrac{18(2\sqrt{2} - 1)}{18}$	Divide top and bottom by 18.
$= 2\sqrt{2} - 1$	
Method 2 $\begin{aligned}12 &= 6 \times \sqrt{2} \times \sqrt{2}\\ &= 2 \times 3 \times \sqrt{2} \times \sqrt{2}\end{aligned}$ Hence $12 - 3\sqrt{2} = 3\sqrt{2}(2\sqrt{2} - 1)$	Factorise $3\sqrt{2}$ out of 12. Show the factorisation of 12 clearly.
Missing side is $2\sqrt{2} - 1$ so: $\begin{aligned}P &= 2 \times 3\sqrt{2} + 2 \times (2\sqrt{2} - 1)\\ &= 6\sqrt{2} + 4\sqrt{2} - 2\end{aligned}$ Perimeter $= 10\sqrt{2} - 2$	Once you have found the missing side show the calculation for getting the perimeter. Remember that the final answer is twice the total of both sides.

2 Algebra

1 Explain the difference between the expressions in each pair.

a $2n$ and $n + 2$ **b** $3(c + 5)$ and $3c + 5$ **c** n^2 and $2n$ **d** $2n^2$ and $(2n)^2$

2 Explain how you know when a letter symbol represents an unknown or a variable.

3 Identify the error in each equation and explain how it should be corrected. Provide some written feedback to the person who did them that does not give them the answer but would help them to see their mistake.

a $5(c + 4) = 5c + 4$ **b** $6(t - 2) = 6t - 4$

c $-3(4 - s) = -12 - 2s$ **d** $15 - (n - 4) = 11 - n$

4 When you substitute $s = 6$ and $t = 2$ into the formula: $z = \dfrac{3(s + 2t)}{12}$ you get 2.5.

Make up two more formulae that also give $z = 2.5$ when $s = 6$ and $t = 2$ are substituted. Explain what you are doing and try to use two different methods as well.

Example 1

An identity can be described as an equation made up of two expressions that mean exactly the same thing. Alternatively, it can be described as an equation that is TRUE no matter the value of the variable. For example:

$\dfrac{a}{4} = a \div 4$ or $2n(n + 4) = 2n^2 + 8n$ or $2^a \times 2^b = 2^{a+b}$

---

5 Write the expression $\dfrac{2n + 6}{2}$ in a different way.

6 The height of an isosceles triangle is three times its base. The area is 6 cm². What is its height?

7 Six friends agreed to buy each other chocolate Easter eggs.

Four of the friends are girls and two of them are boys.

Each girl gives each boy a red egg.

Each boy gives each girl a blue egg.

Each girl gives each of the other girls a yellow egg.

And each boy gives each of the other boys a green egg.

How many eggs of each colour do the friends buy between them?

Make sure you show your working.

8

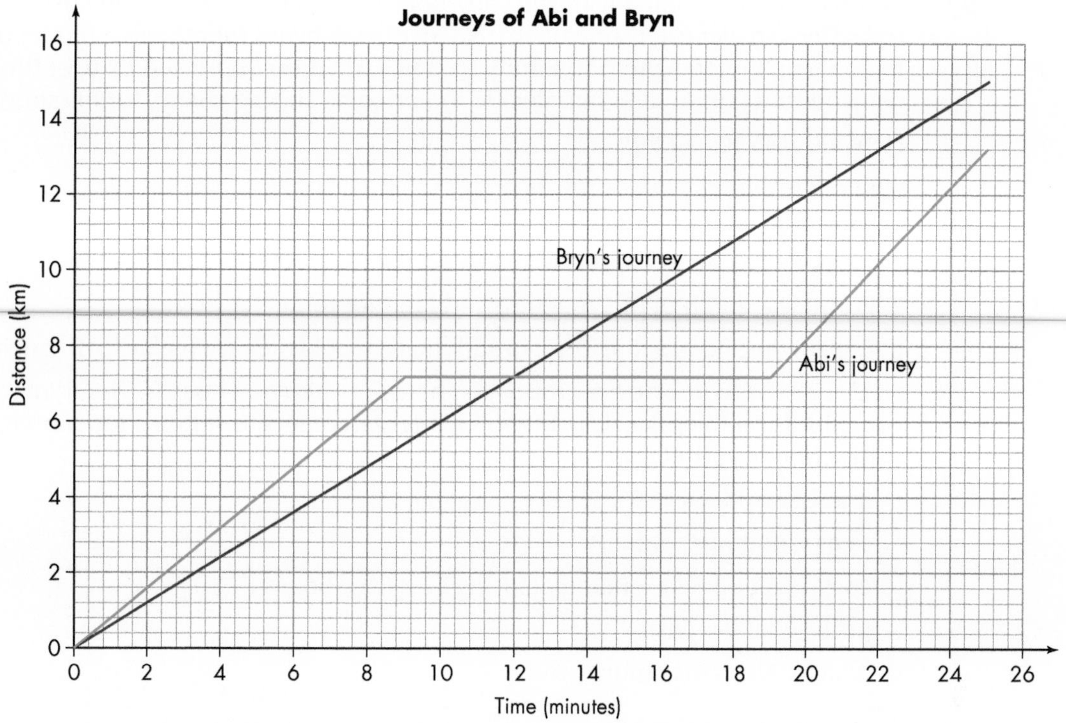

Journeys of Abi and Bryn

The distance–time graph shows the journeys of Abi and Bryn.

a Who was ahead after 5 minutes?

b What happened at 9 minutes?

c After how long did Bryn overtake Abi?

d Who is ahead at 25 minutes and by how much?

e If they both continued at the same speed at which they were travelling at 25 minutes, how much further would Abi need to travel to overtake Bryn?

f Write your own question, similar to this one, that uses the graph of a linear function.

9 When a ball is thrown straight up in the air, its approximate height, h, above the ground is given by the equation $h = ut - 5t^2$, where u is its initial speed and t is the time it is in the air.

a Work out how long the ball is in the air when it is thrown with an initial speed of 16 m/s.

b Estimate the greatest height the ball reaches.

c The equation assumes that the ball starts from the ground. How would you change the equation if you where told that it is released 1 m above the ground?

10 **a** Describe the type of graph that would model the vertical component of the motion of a horse when it jumps over a fence.

You can use the formula $s = ut - \frac{1}{2}gt^2$ to describe this motion, where u is the initial velocity (as the horse starts to jump) and g is the acceleration due to gravity.

b What is the maximum height the horse will reach? Assume $u = 8 \sin \theta$ where θ is the angle the horse makes with the ground on take-off.

c Reflect on your answer. Does your answer make sense?

11 Alex has set up a company to sell snowboards. He has £45 000 set-up costs and it costs £95 to make each snowboard. An inverse demand function is a model of sales that sets up the demand or quantity sold as a function of price. Alex uses this inverse demand function to help him decide the price he should charge for his snowboards.

Demand = 45 000 − 175P

where P is the price of a snow board.

The function includes his set-up costs and a multiplier that includes manufacturing costs as well as other expenses.

a Produce a graph of this function and identify some key points.

b Explain the effect on the formula and what it would mean in reality to his pricing structure.

c Alex needs to decide how much to charge for his snowboards. The number he sells depends on the price. Use the following statements to set up a function f(p) to describe his profit.

number sold = 45000 − 175P

sales = number sold × P

costs = set up fees + manufacturing costs per board

profit = sales − costs

d What is Alex's maximum profit?

12 Look at the list of formulae.

i $V = lwh$ **ii** $s = ut + \frac{1}{2}at^2$ **iii** $x = 3y − 2$ **iv** $s = \left(\dfrac{u + v}{2}\right)t$

v $v^2 = u^2 + 2ad$ **vi** $A = 2\pi r^2 + 2\pi rh$

a Might it be difficult to substitute values into any of them? Which ones? Explain what makes them difficult and identify what the typical mistakes might be.

b Might it be difficult to rearrange any of them? Which ones? Explain what makes them difficult and identify what the typical mistakes might be.

13 Look at the list of formulae.

a $y = 2x + 3$ **b** $z = 2(x + 3)$ **c** $t = −2(3 − x)$ **d** $z = \dfrac{−2(x + 2)}{x}$

Might it be difficult to substitute negative values into any of them? Which ones? Why? What typical mistakes might people make? Make some suggestions to help them avoid making these mistakes.

14 Explain the similarities and differences between rearranging a formula and solving an equation.

15 For homework, a teacher asks her class to simplify the expression $\dfrac{x^2 − 9}{x^2 + 2x − 3}$.

This is Phillip's answer.

$$\dfrac{x^2 − 9}{x^2 + 2x − 3}$$

$$\dfrac{\cancel{x^2} − 9^3}{\cancel{x}(x + 2) − 3_1} = \dfrac{x − 3}{x + 2 − 1}$$

$$= \dfrac{x − 3}{x − 1}$$

Phillip checked the answer and it was correct. However, when the teacher marked the homework, she found that Phillip had in fact made several mistakes.

Explain the mistakes that Phillip made.

16 **a** Explain why: 'I think of a number and double it' is different from 'I think of a number and double it. The answer is 12.'

b $6 = 2p - 8$

    i How many solutions does this equation have?

    ii Write another equation with the same solution.

    iii Why do they have the same solution? How do you know?

17 **a** Imagine you are teaching a short session on expressions, equations, formulae and identities. Explain carefully the similarities and differences between them.

b Produce a short activity that will help your class to become confident and fluent when using expressions, equations, formulae and identities.

18 Without drawing the graphs, compare and contrast features of these pairs of graphs.

    a $y = 2x$ **b** $y = x + 5$ **c** $y = 4x - 5$ **d** $y = 2x$

       $y = 2x + 6$ $y = x - 6$ $y = -4x + 6$ $y = \frac{1}{2}x$

Use as much mathematical vocabulary as possible in your explanations.

19 For real-life problems that generate linear functions explain:

a what the gradient means in terms of the original problem

b whether the intermediate points have any practical meaning

c what the intercept means in terms of the original problem.

20 **a** Explain how can you identify, from its equation:

    i a quadratic function

    ii a cubic function.

b When sketching the graph of a given quadratic function, how do you find an appropriate set of coordinates to base your sketch on?

c Explain why there are no coordinates on the graph of $y = 2x^2 + 5$ that lie below the x-axis. Use two different methods.

21 **a** Explain why a quadratic function has line symmetry.

b Explain why a cubic function does not have line symmetry.

c Describe the symmetry of a cubic function.

22 **a** Match the description to a graph and then write a suitable formula for each.

    i A stone is dropped from a cliff. After 2 seconds it has travelled approximately 20 metres.

    ii Two people can clean a house in 6 hours. How long would it take n people to clean the house?

    iii A taxi firm charges fifty pence a mile with a call out charge of £2.

iv The side length of a square carpet increases as the area of the carpet increases.

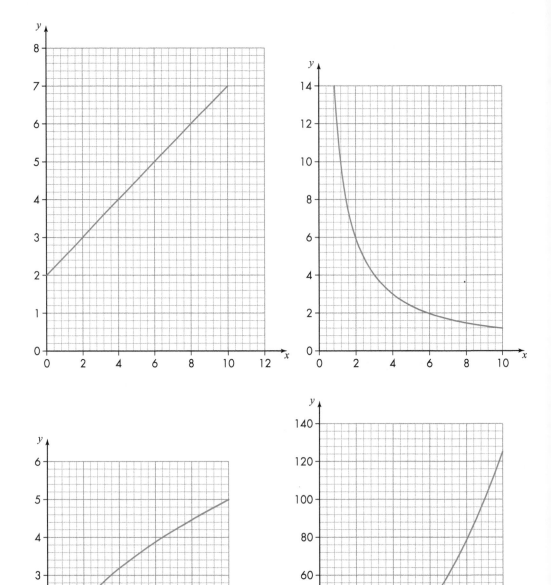

23 A balance is arranged so that 4 kg is placed at 5 units from the pivot on the left-hand side and balanced on the right-hand side by *y* kg placed *x* units from the pivot. Describe how the graph of *y* varies with *x*.

24 **a** Write an example of an equation of a graph that translates the graph of $y = x^3$ upwards. Write another example.

    b What is the same/different about the functions $y = x^2$, $y = 4x^2$, $y = 4x^2 + 3$ and $y = \frac{1}{4}x^2$?

25 Is the following statement always, sometimes or never true?

As a increases the graph of $y = ax^2$ becomes steeper.

26 Write a justification to convince a friend or teacher that the graph of $y = 4x^2$ is a reflection of the graph of $y = -4x^2$ in the x-axis.

27 A motorbike has an initial speed of u m/s. The motorbike accelerates to a speed of $3u$ in 15 seconds. It then travels at a constant speed of $3u$ for 10 seconds. It comes to a stop over the next 20 seconds. How far has the motorbike travelled in terms of u? State any assumptions you make.

28 Points D(5, 0) and E(–5, 0) are the end points of a diameter of a circle.

a What is the equation of the line between these two points?

b Show that the point F(–3, 4) is also on the circumference of the circle.

c Explain how you might find the equation of the tangent at point F.

29 **a** Describe three different methods you can use to solve a pair of simultaneous linear equations.

b Explain how you would decide which method to use. Why might you use more than one method?

30 Use both non-graphical methods to solve this pair of simultaneous equations.

$3x - 4y = 13$

$2x + 3y = 20$

31 Explain the advantages and disadvantages of each method of solving simultaneous equations for each pair of equations.

a $6x + 5y = 23$

$5x + 3y = 18$

b $y = 3x$

$4x + 2y = 5$

c $y - x = 2$

$y = 2x^2 - 8$

32 **a** Is it possible for a pair of linear simultaneous equations to have more than one solution or to have no solution? How do you know? Explain what it means in each case.

b Without drawing the graph or solving the pairs simultaneous equations, explain how you know how many solutions there are just by looking at them.

i $y - 2x = -5$

$y = 0.5x + 1$

ii $2y - 4x = 10$

$3y - 6x = 18$

iii $2y - 4x = -10$

$y = 2x - 5$

c Now solve each pair of equations. Explain how this supports your answer to part **b**.

d How does a graphical representation help you with the number of solutions?

33 **a** Tommy is solving the simultaneous equations:

$5x - y = 9$ and $15x - 3y = 27$.

He finds a solution of $x = 2$, $y = 1$. Explain why this is not a unique solution.

b Laura is solving the simultaneous equations:

$5x - y = 9$ and $15x - 3y = 18$.

Explain why is it impossible to find a solution.

34 **a** Solve the simultaneous equations $y = x^2 + 2x - 5$ and $y = 6x - 9$.

b Which sketch represents the graph of the equations in part **a**? Explain your choice, giving mathematical reasons.

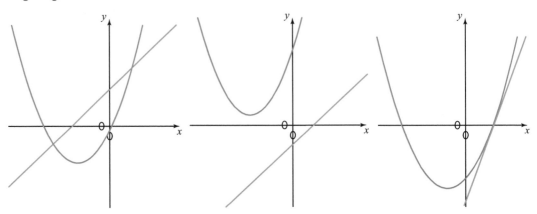

35 **a** Ten second-class and six first-class stamps cost £9.02. Eight second-class and 10 first-class stamps cost £10.44. How much do I pay for three second-class and four first-class stamps?

b Henri pays £4.37 for six cans of cola and five chocolate bars. On his next visit to the shop he pays £2 for three cans of cola and two chocolate bars. A few days later, he buys two cans of cola and a chocolate bar. How much do they cost him?

36 A sales manager compares the commission of three sales assistants. He finds that:

• Sarah earned 25% more commission than Benjamin

• Benjamin's commission was $\frac{2}{5}$ of Charlotte's

• the total of Sarah's and Charlotte's commission was £75.

a How much commission did each sales assistant earn?

b What method did you use? Explain how you checked your answer.

37 **a** What method could you use to solve these problems?

i A number plus its cube is 20. What is the number?

ii The length of a rectangle is 2 cm greater than the width. The area is 67.89 cm². What is the width?

b Use your method to solve both problems.

38 A smaller square is inscribed inside a larger square as shown in the diagram. Prove that less than half the larger square is shaded.

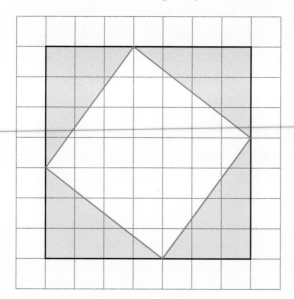

39 The next term of a sequence is found by multiplying the previous term by a and then subtracting b. a and b are positive whole numbers.

Explain how you can use this information to work out the next two terms in the sequence 5, 23, 113, ..., ...

40 **a** The sum of the whole numbers from 1 to 50 is 1275. Use this to work out the sum of the whole numbers from 2 to 51.

b What could you do to help you check your method?

c The formula for the sum of the first n natural numbers is:

$$S_n = \frac{n(n + 1)}{2}$$

Use this to check your answer to part **b**. If it is wrong, work out why.

d Johann Carl Friedrich Gauss was a famous German mathematician born in 1777. When Gauss was still in primary school, he was asked to find the sum of the natural numbers from 1 to 100. He found the answer quickly by finding a pattern which led to the formula you used in part **c**. Gauss realised that he could add all the numbers in pairs to get 101 in each case.

1 + 100 = 101

2 + 99 = 101

3 + 98 = 101

...

48 + 53 = 101

49 + 52 = 101

50 + 51 = 101

Use this to explain why the formula in part **c** works.

41 Ali receives £1000 in year 1. He receives £2000 in year 2, £3000 in year 3 and so on for 20 years.

Ben receives £1 in year 1. He receives £2 in year 2, £4 in year 3 and so on.

a Show that the formula $500n(n + 1)$ gives the total amount that Ali has after n years.

b Show that the formula $2^n - 1$ gives the total amount that Ben has after n years.

c Produce a brief report for Ali and Ben to explain the outcomes for them over 20 years.

42 A sequence of fractions is $\frac{3}{7}, \frac{5}{10}, \frac{7}{13}, \frac{9}{16}, \frac{11}{19}, \cdots$

a Write down an expression for the nth term of the numerator.

b Write down an expression for the nth term of the denominator.

c Work out the fraction when $n = 1000$.

d Will the terms of the series ever be greater than $\frac{2}{3}$? Explain your answer.

43 **a** Look at these sequences. One is arithmetic, one is geometric and one is neither. Identify which is which. Justify your decision by describing the rule that generates the sequence.

   i 1, 1, 2, 3, 5, 8, 13, …

   ii 3, 6, 12, 24, 48, 96, …

   iii –3, 1, 5, 9, 13, 17, …

b Look at the rules you generated in parts **a** and **b** in Q42. What type of sequence do these rules generate?

c What type of sequence is the overall expression you used in part **c** of Q42?

44 You can write an arithmetic sequence in the form:

$a, a + d, a + 2d, a + 3d, a + 4d, \ldots$

a What do the letters a and d stand for?

You can generate any term in this sequence, using the formula $X_n = a + (n - 1)d$.

b Check this, using the example of an arithmetic sequence in part **a**.

c You can write a geometric sequence in the form: $a, ar, ar^2, ar^3, ar^4, \ldots$

Write a general formula similar to the one in part **b** to generate the terms in the sequence.

d In Q40 you worked out a formula for the sum of the first n whole numbers. What type of sequence is this? Explain how you know.

45 Here is a proof of the formula to calculate the sum of a arithmetic sequence.

Proof of the sum of an arithmetic sequence

Write the sum of the first n terms of an arithmetic sequence S_n.

$S_n = a + (a + d) + (a + 2d) + (a + 3d) + \dots + (a + (n - 1)d)$

Now write the same sum in reverse order.

$S_n = (a + (n - 1)d) + \dots + (a + 3d) + (a + 2d) + (a + d) + a$

Add corresponding terms (first terms, second terms, third terms, etc.) together to get $2S_n$.

S_n	$a+$	$a+d+$	$a+2d+$	$a+3d+$	$+...+$	$a+(n-1)d$
S_n	$a+(n-1)d$	$a+(n-2)d$	$a+(n-3)d$	$a+(n-4)d$	$+...+$	a
$2S_n$	$2a+(n-1)d$	$2a+(n-1)d$	$2a+(n-1)d$	$2a+(n-1)d$	$+...+$	$2a+(n-1)d$

Each term is $2a + (n - 1)d$ and there are n of them.

Therefore $2S_n = n(2a + (n - 1)d)$

$S_n = \dfrac{n}{2}(2a + (n - 1)d)$

a Work through the proof carefully and then try reproducing the proof without referring to it.

b Show that this formula generates the same answer as you got for the sum of the first n numbers in Q41.

46 You are now going to complete a similar proof for the sum of a geometric sequence.

a Write the first n terms of the geometric sequence

$S_n = \dots$

b Multiply S_n by r.

$rS_n = \dots$

c Subtract rS_n from S_n. Use the table below to help you.

S_n		+	+	+...+	+	
rS_n		–	–		–	–
$S_n - rS_n$						

Therefore:

$S_n - rS_n = \dots$

d Factorise the equation.

e Rearrange the equation.

47 Show that $(x + 1)^2 = x^2 + 2x + 1$ by considering the area of an $(x + 1)$ by $(x + 1)$ square.

48 a Write an example of an inequality. Now write a set of instructions for finding the solution set for your inequality.

b What are the important conventions when representing the solution set of an inequality on a number line?

c Explain why the inequality sign changes when you multiply or divide an inequality by a negative number.

49 a Provide an example to demonstrate why you need a minimum of three linear inequalities to describe a closed region.

b Explain how you check if a point lies:

i inside the region

ii outside the region

iii on the boundary of the region.

50 Albion Rovers are at the top of a football league. They have played all their games and have 58 points. The only other team that could come top of the league are Albion United. They currently have 50 points, with four games still to play. Teams are awarded three points for a win, one point for a draw and no points for a loss.

Alice is a keen Albion United fan and she draws this graph to show the possible outcomes.

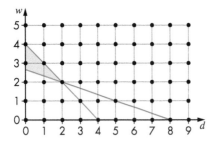

She also writes:

$w + d \leqslant 4$

$3w + d \geqslant 8$

She says that the shaded region shows what Albion United need to do to become league champions.

a Explain what Alice has done and how to use the graph.

b Describe in words what Albion United need to do to win the league.

c How would the graph be different if Albion United had five games still to play instead of four?

51 a Give an example of a quadratic sequence.

b Explain the similarities and differences between the way you find the nth term for a linear sequence and the way you find the nth term for a quadratic sequence.

c Continue the sequence that starts with 2, 8, ...:

• so that it is a linear sequence

• in a different way so that it is a quadratic sequence.

Explain how you did this for both examples, referring to the general rules for deciding whether the nth term of a sequence is linear or quadratic.

52 $(m^2 - n^2)$, $2mn$ and $(m^2 + n^2)$ are the sides of a right-angled triangle where m and n are integers.

For example, let $m = 5$ and $n = 3$.

$m^2 - n^2 = 16$

$2mn = 30$

$m^2 + n^2 = 34$

$34^2 = 1156$

$16^2 + 30^2 = 256 + 900 = 1156$

Prove this result algebraically.

53 Explain why the triangular number sequence 1, 3, 6, 10, 15, 21, 28, ... follows the pattern of two odd numbers followed by two even numbers.

54 $10p + q$ is a multiple of 7. Prove that $3p + q$ is also a multiple of 7.

55 Show that $2(5(x - 2) + y) = 10(x - 1) + 2y - 10$.

56 Prove why these calculations work.

a Think of two numbers less than 10.

Subtract 2 from the larger number and then multiply by 5.

Add the smaller number and multiply by 2.

Add 9 and subtract the smaller number.

Add 1 to both the tens digits and the units digits to obtain the numbers first thought of.

b Choose two numbers: one with one digit, the other with two digits.

Subtract 9 times the first number from 10 times the second number.

The units digit of the answer is the single-digit number chosen. Split the answer into two numbers – the last digit and the other two digits. The sum of these two numbers is the other number chosen.

57 Prove that: $(n - 1)^2 + n^2 + (n + 1)^2 = 3n^2 + 2$.

58 What is the nth term of the sequence 4, 9, 14, 19, 24, ...?

59 When T_n represents the nth triangular number, prove that: $T_n = \frac{1}{2}n(n + 1)$.

60 When T_n represents the nth triangular number, prove that: $3T_n = T_{2n+1} - T_{n+1}$.

61 When T_n represents the nth triangular number, prove that:

$$\frac{T_n - 1}{T_n} = \frac{(n - 1)(n + 2)}{n(n + 1)}$$

62 Prove that the sum of any five consecutive numbers in a sequence is a multiple of 5.

63 $10^x = \dfrac{a}{b}$, $10^y = \dfrac{b}{a}$

Prove that $x + y = 0$.

64 Prove that the square of the sum of two consecutive integers minus the sum of the squares of the two integers is four times a triangular number.

65 p, q and r are three consecutive numbers. Prove that $pr = q^2 - 1$.

66 **a** Write an expression that is equivalent to $\frac{z}{2} - q(q+1) - 4$.

  b Write an expression that would simplify to:

  i $\frac{3x + 2y}{5}$ **ii** $5x\left(\frac{4x}{5} + 2\right)$.

67 **a** Explain what it means to factorise a quadratic expression.

  b Explain what information you need to complete this.

  c Explain what it tells you about the factors if the constant term of the quadratic is negative.

  d Explain what difference it makes if the constant term is zero.

68 **a** Write an expression that can be written as the difference of two squares. Explain how you know it can be written as the difference of two squares.

  b Why must 1000×998 give the same result as $999^2 - 1$?

69 Give two examples of algebraic fractions that can be cancelled and two that cannot be cancelled. Explain how you decided on your examples.

70 An expression of the form $\dfrac{ax^2 - b}{cx^2 + dx^2 - e}$ simplifies to $\dfrac{3x + 4}{x + 2}$.

  What was the original expression?

71 Explain how the product of two linear expressions of the form $(2a \pm b)$ is different from the product of two linear expressions of the form $(a \pm b)$.

72 **a** Use the sine rule and the general formula for the area of a triangle to show that the area of any triangle is equal to $\frac{1}{2}ab\sin C$.

  b Show that when $x = 4$ the area of this triangle is equal to $3\sqrt{2}$.

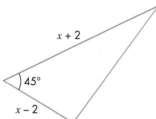

$x + 2$

$45°$

$x - 2$

73 The functions f and g are such that:

  $f(x) = 3 - 7x$ $g(x) = 7x + 3$.

  Prove that $f^{-1}(x) + g^{-1}(x) = 0$ for all values of x.

74　**a** What is special about two linear expressions that, when multiplied together, have:

　　i a positive x coefficient

　　ii a negative x coefficient

　　iii no x coefficient?

b Give an example of an expression in the form $(x + a)(x + b)$ that, when expanded, has:

　　i the x coefficient equal to the constant term

　　ii the x coefficient greater than the constant term.

c What does the sign of the constant term tell you about the original expression?

Example 2

What number should you add to $x^2 + 3x$ to complete the square?

The general formula for completing the square is $x^2 + bx + \left(\dfrac{b}{2}\right)^2$.

So in this case you will need to add $\left(\dfrac{3}{2}\right)^2 = \dfrac{9}{4}$.

75　**a** What number should be added to $x^2 + 5x$ to complete the square?

b Solve the quadratic equation $2x^2 + 10x - 5 = 0$ by completing the square.

c Adam tried to solve a quadratic by completing the square but he made a number of mistakes. Look at his working. What mistakes did he make?

$x^2 + 5x - 5 = 0$

$$\left(x + \frac{5}{2}\right)^2 = \left(\frac{5}{2}\right)^2$$

$$\left(x + \frac{5}{2}\right)^2 = \frac{25}{2}$$

$$x = \frac{5}{2} + 5\sqrt{2}$$

$$x + \frac{5}{2} = \sqrt{\frac{25}{2}}$$

d Copy his working and write some comments that show Adam where he has gone wrong. Write them in a way that does not correct his answer but will help him not to make the same mistake again. Your answer should include guidance on why he should have known he had made a mistake this time as well as how to make sure he doesn't make the same mistake again.

76 For each part of this question think of a quadratic equation of the type $y = (x \pm a)^2 \pm b$. Give an example of a quadratic that fits the description. Justify your example in each case.

   a The turning point has a positive x-value.

   b The turning point has a positive y-value.

   c The y-intercept is positive.

77   **a** Sketch the graph of $f(x) = x^2 + x + \dfrac{9}{4}$.

   b Hence determine whether $f(x + 3) - 2 = 0$ has any real roots. Give a clear justification for your answer.

⑦Hints and tips

1, 2	You may find a couple of examples will help you explain this.
6	You may find it useful to draw a diagram and make sure you read the question carefully.
9a	At the start and at the end, $h = 0$. Substitute $h = 0$ and $u = 16$ into the original equation and solve it.
9b	The greatest height reached occurs halfway through the time the ball is in the air.
10b	You will need to make an assumption about the angle at which the horse takes off and how long it is in the air, to model this. Drawing a graph may help you see what you need to do. Use completing the square to identify key points in the horse's trajectory.
11a	Think carefully what 'minimum demand' means in terms of the equation. This formula would be called a demand curve.
14	You may find a couple of examples will help you explain this.
17	You may want to use some examples to support your explanation.
19	It might help your explanation if you use some real-life examples.
21, 22	Sketch a graph.
24b	Use $y = x^2$ as your starting point and consider what happens to the graph in each case.
27	Sketch a graph. The distance is area under the graph.
28a	Draw a diagram.
28c	The equation for a tangent at point (x_0, y_0) is $y - y_0 = -\dfrac{x_0}{y_0}(x - x_0)$.
30	Remember, you should never get one of these questions wrong as you can always substitute your values for x back into the original equations. This type of self-checking is true for many types of question involving algebra.
32	Sketch and describe the graphs in each case.
33b	You may want to refer to what you did in Q32.
40b	Try a simpler example, such as the sum of the numbers from 1 to 6.
41	You could draw a graph showing the key points.
44c	$X_n = ar^{\cdots}$

46c	Here is a partially completed table.

S_n	a	$+ ar$	$+ ar^2$	$+ \dots +$	$+ ar^{n-1}$	
rS_n		$- ar$	$-$		$+ ar^{n-1}$	$+ ar$
$S_n - rS_n$	a	$ar - ar$				

46d	$S_n(1 - r) = \dots$ and rearrange.

47	Draw a square.

51b	You may want to use some examples to help explain your answer but remember you need more than an example to justify your response.
56b	For example, choose 7 and 23. $(23 \times 10) - (7 \times 9) = 167$. The single-digit number chosen is 7, the two-digit number chosen is $16 + 7 = 23$.
60	For example, $T_4 = 10$, $T_9 = 45$, $T_5 = 15$; $3 \times 10 = 45 - 15$.
64	For example, let the two integers be 6 and 7. $(6 + 7)^2 - (6^2 + 7^2) = 169 - 85 = 84 = 4 \times 21$.
67	Explain how you work out the two linear factors.
72	First use what you did in part **a** to show that the area of this triangle can be written as $\frac{\sqrt{2}}{4}(x^2 - 4)$. Also, $\sin 45°$ is equal to $\frac{\sqrt{2}}{2}$.
74	In each case consider what it means when you expand two linear functions of the type $(ax + b)(bx + c)$.
76	Give an explanation as well as an example.

Worked exemplars

 1 **a** Expand $(5x + 3y)(3x - 4y)$.

   b Use your answer to find the value of 50.03×29.96 without using a calculator.

> This is a question requiring mathematical reasoning. This means that once you have expanded the brackets for part **a**, you then need to find a link between parts **a** and **b** so that you can use your answer from **a** to help you with **b**.

a $(5x + 3y)(3x - 4y) = 15x^2 - 20xy + 9xy - 12y^2$ $\qquad\qquad\qquad\quad = 15x^2 - 11xy - 12y^2$	
b $x = 10$ $\quad y = 0.01$	Comparing the algebra in **a** with the multiplication sum in **b**, you need to recognise that they are the same when $x = 10$ and $y = 0.01$.
$15x^2 - 11xy - 12y^2$ $= 15 \times 10^2 - 11 \times 10 \times 0.01 - 12 \times 0.01^2$	Substitute $x = 10$ and $y = 0.01$ into the expansion.
$= 1500 - 1.1 - 0.0012$ $= 1498.8988$	

 2 A bookshelf holds P paperback and H hardback books. The bookshelf can hold a total of 400 books. Explain which of the following may be true.

   a $P + H < 300$

   b $P \geqslant H$

   c $P + H > 500$

> This is a question on communicating mathematics so you need to make sure you communicate information accurately and articulately.

a The first inequality ($P + H < 300$) could be true. The bookshelf doesn't have to be full.	Although it is important to state correctly whether or not a given statement is true, the emphasis is on the quality of your explanation.
b The second inequality ($P \geqslant H$) could be true. There could be more paperbacks than hardbacks.	
c The third inequality ($P + H > 500$) cannot be true. There can only be a maximum of 400 books.	

3 The graph of $y = x^2 - 3x + 1$ has been plotted.

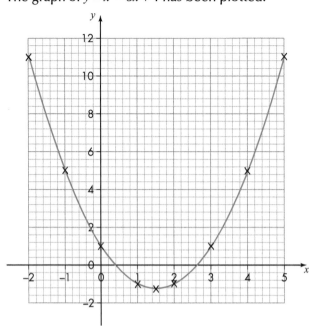

a By plotting an appropriate straight line on the same graph, solve the equation
$x^2 - 4x - 2 = 0$.

b Comment on the accuracy of your answers.

This is an evaluation question, which means that you need to consider and analyse your results.	
a $y = x^2 - 3x + 1$ $\dfrac{0 = x^2 - 4x - 2}{y = \qquad x + 3}$	Start by determining the equation of the straight line that needs to be plotted.
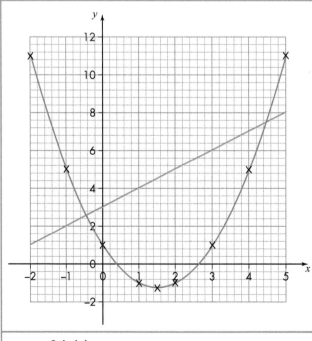	Plot $y = x + 3$.
$x = -0.4, 4.4$	Read off x values from intersection points correct to 1 decimal place.

b Quadratic formula: $a = 1$, $b = -4$, $c = -2$	To check the accuracy of your results, use another method to solve the equation.
$x = \dfrac{-b \pm \sqrt{b^2 - 4ac}}{2a}$ $= \dfrac{-(-4) \pm \sqrt{(-4)^2 - 4(1)(-2)}}{2(1)}$ $= \dfrac{4 \pm \sqrt{24}}{2} = -0.449\,489\,7, \ 4.449\,489\,7$ (7 dp)	Use the quadratic formula (or complete the square) to solve $x^2 - 4x - 2 = 0$. Write the answers to a greater accuracy than you found by the graphical method.
Or: Completing the square: $(x - 2)^2 - 4 - 2 = 0$ $(x - 2)^2 = 6$ $x - 2 = \pm\sqrt{6}$ $x = 2 \pm \sqrt{6} = -0.449\,489\,7, \ 4.449\,489\,7$ (7 dp)	
Answers found by the method of intersection were accurate to 1 decimal place, but algebraic methods were more accurate.	Finally, evaluate the methods used and results obtained.

(MR) **4** The sketch shows the graph of $y = x^2 - 6x + 5$.

The minimum point of the graph is $(3, -4)$.

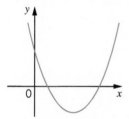

a Describe what happens to the graph of $f(x) = x^2$ under the transformation $f(x - 3)$.

b Describe what happens to the graph of $f(x) = x^2$ under the transformation $f(x) - 4$.

c Explain how the answers to **a** and **b** connect the equation $y = x^2 - 6x + 5$ and the minimum point $(3, -4)$.

a A translation of $\begin{pmatrix} 3 \\ 0 \end{pmatrix}$ **b** A translation of $\begin{pmatrix} 0 \\ -4 \end{pmatrix}$	For parts **a** and **b**, write the transformations as translations, using column vectors. Remember that the top number is a move in the x-direction and the bottom number is a move in the y-direction.
	For part **a**, the transformation translates the graph 3 units to the right.
	For part **b**, the transformation translates the graph 4 units downwards.
c $x^2 - 6x + 5 = (x - 3)^2 - 4$ This means that the graph function is a translation of x^2 by $\begin{pmatrix} 3 \\ -4 \end{pmatrix}$. This takes the original minimum point $(0, 0)$ to $(3, -4)$.	You have to use mathematical reasoning in part **c**. Remember that writing the equation in 'completing the square' form $(x - a)^2 - b$ gives the minimum point $(a, -b)$. The transformations $f(x - 3)$ and $f(x) - 4$ combined transform x^2 into $(x - 3)^2 - 4$, which is the same as $x^2 - 6x + 5$.

5 Given that $f(x) = \dfrac{x^2 + 3x - 10}{2x^2 - 9x + 10}$, prove that $f^{-1}(3) = 4$.

This is a problem-solving question so you need to plan a strategy to solve it and, most importantly, communicate your method clearly. You need to show each step clearly.

There are two different methods shown here.

Method 1	Factorise and cancel the numerator and denominator.
$f(x) = \dfrac{x^2 + 3x - 10}{2x^2 - 9x + 10}$ $\quad\quad = \dfrac{(x + 5)(x - 2)}{(2x - 5)(x - 2)}$ $f(x) = \dfrac{x + 5}{2x - 5}$	
$y = \dfrac{x + 5}{2x - 5}$ $y(2x - 5) = x + 5$ $2xy - 5y = x + 5$ $2xy - x = 5y + 5$ $x(2y - 1) = 5y + 5$ $x = \dfrac{5y + 5}{2y - 1}$ $f^{-1}(x) = \dfrac{5x + 5}{2x - 1}$	Find the inverse function.
$f^{-1}(3) = \dfrac{5 \times 3 + 5}{2 \times 3 - 1} = \dfrac{20}{5} = 4$	Substitute $x = 3$.
Method 2 $3 = \dfrac{x^2 + 3x - 10}{2x^2 - 9x + 10}$	Put the function equal to 3.
$3(2x^2 - 9x + 10) = x^2 + 3x - 10$	Multiply by the denominator.
$6x^2 - 27x + 30 = x^2 + 3x - 10$ $5x^2 - 30x + 40 = 0$ $x^2 - 6x + 8 = 0$	Simplify.
$(x - 2)(x - 4) = 0$ $x = 2$ or 4	Factorise.
Not $x = 2$ because $f(x)$ is undefined for $x = 2$ in its original form, since both the numerator and denominator would equal zero, and division by zero is forbidden. $x = 4$, so $f^{-1}(3) = 4$	Explain why $x = 2$ would not be allowed.

3 Ratio, proportion and rates of change

1 **a** Is the ratio 1 : 6 the same as the ratio 6 : 1? Explain your answer.

    b Show that 19 : 95 is the same ratio as 1 : 5.

    c The instructions on a packet of cement say: 'mix sand and cement in the ratio 6 : 1.' A builder mixes 6 kg of sand with one bucketful of cement.

       Could this be correct? Explain your answer.

    d The ratio of boys to girls at a school club is 2 : 5.

       Could there be 24 students at the club altogether? Explain your answer.

2 Batteries are sold in packs.

A pack of 3 batteries costs £1.50.

A pack of 15 batteries costs £5.

A pack of 25 batteries costs £10.

    a Work out the cheapest way to buy **exactly** 90 batteries.

    b The packs of 15 are now available for 'buy two get one free'. Does this change your answer?

3 The answer to a percentage increase question is £10.

    a Make up an easy question.

    b Make up a difficult question.

       Explain what makes the questions easy or difficult.

4 Two metal objects are the same size and look identical, but they have different masses.

Explain how this tells you that they are probably made from different metals.

5 A storage area has 30 tonnes of sandstone.

The density of sandstone is 2.3 g/cm^3.

    a What is the volume of sandstone in the storage area?

       Give your answer in m^3.

    b The density of granite is 2.7 g/cm^3.

       The same volume of granite is stored as the volume of sandstone.

       How much heavier is the granite?

6 The density of a piece of oak is 630 kg/m^3.

The density of a piece of mahogany is 550 kg/m^3.

Two identical carvings are made, one from oak and the other from mahogany.

The oak carving has a mass of 315 grams.

What is the mass of the mahogany carving?

7 **a** A golf club has 24 women members. The ratio of men to women members of a golf club is 5 : 2. How many members does the golf club have?

b Richard, Shaun and Joan are paying a restaurant bill of £85. They want to split the bill in the ratio 2 : 3 : 5. How much does Shaun pay?

c Design your own question like the one in part **a**. Try to use a different context if you can.

8 **a** Sarah is comparing tablet prices. The battery life of the more expensive tablet is $\frac{5}{4}$ times the battery life of the cheaper tablet.

The cheaper tablet has 8 hours battery life. How long does the more expensive tablet have?

b The more expensive tablet costs £198. The cheaper one costs £118. Is this less than or greater than the proportional change to the battery life? Justify your answer.

c Sarah thinks she may be able to negotiate a reduction. What reduction would she need to negotiate to make it worth buying the more expensive model, based on the proportional change in battery life?

9 Sally wants to buy some blank CD-R discs. She is trying to decide which deal is best.

5 – Pack	10 – Pack	5 – Pack
90 minutes each	80 minutes each	80 minutes each
£6.50	£6.50	£4.00

a Which pack of CDs is the best buy?

b Why might someone not choose the best buy?

10 Mr and Mrs Fitzpatrick are going to Germany. They each have £400 to change into euros. They see this deal in a bank.

> **Fantastic Rates**
>
> Get 1.19 Euros for £1
> on amounts less than £500
>
> Get 1.22 Euros for £1
> on amounts greater than £500

How much more money will they get by putting their money together before they change it?

11 **a** Justify these ideas.

i Dividing by $\frac{1}{2}$ is the same as multiplying by 2.

ii Dividing by $\frac{1}{3}$ is the same as multiplying by 3.

iii Dividing by $\frac{2}{3}$ is the same as multiplying by $\frac{3}{2}$.

b Use this to explain how to divide by fractions.

12 Philippa makes biscuits that she sells at children's parties.
She has these ingredients.

8 kg plain white flour

3 kg caster sugar

2 kg butter

7 kg icing sugar.

To make 24 birthday biscuits she needs:

    250 g plain white flour

    85 g caster sugar

    20 g unsalted butter

    2 tbsp lemon curd

    250 g icing sugar

    1 tbsp strawberry jam

She has plenty of jam and lemon curd.

a Philippa sells her biscuits in packs of 15.

How many packs of biscuits can she make from the ingredients she has?

b Philippa sells three-quarters of her biscuits to individuals at £2.99 per pack.

The rest she sells in bulk to one buyer at a discount of 15%.

The ingredients cost her £59 plus an additional £26 for extras such as packing.

What is her percentage profit?

13 John is buying a new computer which is advertised for £595 exclusive of VAT.

VAT is charged at 20%.

John says: 'I managed to negotiate a 20% discount so I only need to pay the shop £595.'

Do you agree or disagree with this statement? If he pays the shop £595 will he have got the deal he thinks he has?

Explain your response to John.

14 **a** The original value before a sale was £A. The sale reduction was 15%.

Write a formula to describe this percentage change.

b Now rearrange the formula to show how you work out the original value when you know the sale price.

c Can you use this approach for all percentage change questions?

Explain your answer.

d Write a similar question, but in a different context, where you can use your formula.

15 Explain your answer for each part of this question.

   a Is a 50% increase followed by a 50% increase the same as doubling?

   b Which deal is a better one? An 80% discount or a 60% discount and a further 20% special offer discount?

   c A shop is offering a 25% discount. Is it better to have this before or after VAT is added at 20%?

16 **a** After one year the value of a scooter has depreciated by $\frac{1}{7}$ and is valued at £996. What was its value at the beginning of the year?

   b Waiting staff at a restaurant get a 4% wage increase.

     The new hourly rate is £6.50.

     What was it before the increase?

   c Sadie has savings of £957.65 after 7% interest has been added.

     What was the original amount of her savings before the interest was added?

   d How do you find a multiplier to calculate an original value after a percentage increase or decrease?

   e How can you tell whether a multiplier increases or a decreases a quantity?

17 Emilia is a sales person for a makeup company.

Her monthly salary is a combination of a fixed monthly salary plus commission on her sales.

Her commission is a fixed percentage of the sales she makes that month.

The table below gives some information on her salary over a 3-month period.

Month	Monthly salary	Sales
April	£1544	£24 000
May	£1568	£28 000
June	£1562	£27 000

   a In July Emilia's monthly take-home salary was £1553.

     What was her sales figure, in pounds, for July?

   b Design your own question like the one in part **a**. Try to use a different context.

18 Twice as many people visit a shopping centre on Saturdays than on Fridays.

The number visiting on both days increases by 50% in the week before Christmas.

How many more visit on this Saturday than on this Friday?

Give your answer as a percentage.

19 The time taken to build an extension is inversely proportional to the number of workers.

It takes 2 workers 20 working days to complete an extension.

   a Three workers start an extension on Monday morning.

     When will they complete it? Show your working. Assume they work 5 days per week.

   b Give a reason why the time taken might not be inversely proportional to the number of workers when the number of workers is very large.

20 A taxi company has a fares structure based on three things: a minimum fare of £2.50, a charge for the time taken and a charge for the distance travelled.

The charge for the time taken is directly proportional to the time taken.

The charge for the distance travelled is directly proportional to the distance travelled.

You have collected this information about the company's fares.

Time taken	2 minutes	5 minutes	10 minutes	12 minutes	15 minutes
Distance	1 mile	2 miles	3 miles	5 miles	6 miles
Total charge	£2.50 (minimum fare)	£4.00	£6.50	£9.90	£12.00

Use this information to work out the charge per mile and the charge per minute, then suggest a competitive pricing structure for another taxi firm.

21 **a** Explain why travelling a distance of 30 miles in 45 minutes gives an average speed of 40 mph.

b Explain what might be a common mistake people make when answering part **a** of this question.

c Explain how the units of speed help you to solve a problem.

You can use your answer to part **a** to help you explain this.

d Make up some easy questions that involve calculating speed, distance or time.

Make up some difficult questions.

What makes them difficult?

22 Don wants to do some planting in his garden. He marks out a shape with an area of 2 m². However, he then decides that he wants to use more garden space.

What is the area of a similar shape in which the lengths are four times the corresponding lengths of the first shape?

23 A can of paint is 30 cm high and holds 5 litres of paint.

How much paint does a similar can that is 75 cm high hold?

24 A sculptor has made a model statue that is 15 cm high and has a volume of 450 cm³.

The real statue will be 4.5 m high. In order to buy enough materials, she needs to know the volume of the real statue.

Work this out for her, giving your answer in m³.

25 Tim has a large tin full of paint that he wants to empty into a number of smaller tins.

The diagram shows the two sizes of tins. The tins are similar.

How many small tins can he fill from one large tin?

26 **a** The length of each side of a square is increased by 15%. By what percentage does the area increase?

b The length of a rectangle is increased by 15%. The width is decreased by 5%.

By what percentage does the area change?

27 The numbers in sets A, B and C are in direct proportion.

A	B	C
5		17
		13.6
2	1.6	
	6.4	
12		
		23.12
	2.24	

a Is there sufficient information to find all the missing entries? Explain your answer.

b What is the maximum number of items that could be entered so that the task remains impossible?

c What is the minimum number of entries needed and what is important about their location?

d What is the best starting point for any empty cell? How many different starting points are there?

28 Sam invests £8000 in an account for 2 years. The account pays 2.7% compound interest per annum.

Sam has to pay 20% tax on the interest earned each year. The tax is taken from the account at the end of the year.

Sam calculates that, at the end of two years, she will have at last £8500 in this account. Is she correct? Explain your answer.

29 Each week, Raja takes out 20% of the amount in his bank account to spend. After how many weeks will the amount in his bank account have halved from the original amount?

30 Which graph matches each function? Explain your answer in each case.

i $f(x) \propto x^2$ **ii** $f(x) = 2x, x > 0$ **iii** $f(x) \propto x$
                                $= -2x, x < 0$

iv $f(x) \propto \dfrac{1}{x^2}$ **v** $f(x) \propto -x$ **vi** $f(x) \propto \dfrac{1}{x}$

a

b

c

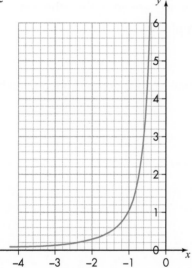

d

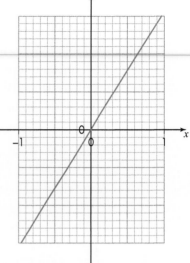

e

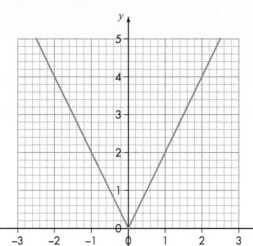

f

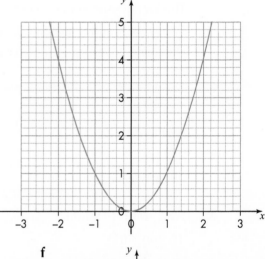

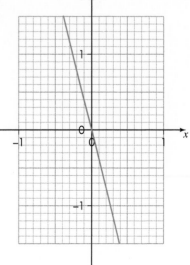

31 **a** Explain to your friend what inverse proportion means.

     b The general formula for inverse proportion is $y \propto \dfrac{1}{x}$.

       Rewrite it as an equation that you can use to solve a problem.

     c Write a problem that requires you to use the equation that you wrote in part **b** again.

32 Newton's law of universal gravitation states that the force (F_g) between two objects is inversely proportional to the square of the distance between the two objects multiplied by the product of the masses (m_1 and m_2) of the two objects.

     The constant of proportionality (G) in the equation is known as the universal gravitational constant.

     a Write an algebraic function to describe this relationship.

     b The radius of the Earth is approximately 6000 km.

     A passenger is on an aeroplane that takes off from sea level and rises to a maximum height of 12 km.

     What is the scale factor that converts the force due to gravity on the passenger just before take off to the force due to gravity at the maximum height?

     c Why is the passenger unlikely to notice the difference?

33 Two cars are 30 miles apart but travelling towards each other.

     The average speed of one car is twice the average speed of the other car.

     The slower car has an average speed of 20 mph.

     How long is it before the two cars meet?

34 The gradient of a chord is equal to the average (mean) of the values of f(x) between x_1 and x_2.

$$\text{Gradient} = \frac{\text{f}(x_2) - \text{f}(x_1)}{2}$$

     What is the gradient of the chord between the points where $x = 2$ and $x = 4$ on the graph of $4y = 2x^2$?

35 Look at this function.

$$\frac{\text{f}(x + h) - \text{f}(x)}{h}$$

     Show that for f(x) = $\dfrac{x^2}{2}$ this function gives the same answer for the gradient as question 34 does.

36 When h gets very small, the function

$$\frac{f(x + h) - f(x)}{h}$$

tends towards the gradient of the tangent to the curve.

For example, for the function $f(x) = x^2$, at $x = 5$:

$$\text{gradient} = \frac{(x + h)^2 - x^2}{h}$$

$$= \frac{x^2 + 2xh + h^2 - x^2}{h}$$

$$= 2x + h$$

As h gets small this equals $2x$ so the gradient at 5 is 10.

a Prove that this formula works for a straight line $f(x) = mx + c$.

b The diagram shows the graph of the function $f(x) = \frac{x^2}{2}$ with the gradient shown at the point where $x = 2$.

Show that the formula for the gradient of the tangent gives the same solution as can be calculated from the graph.

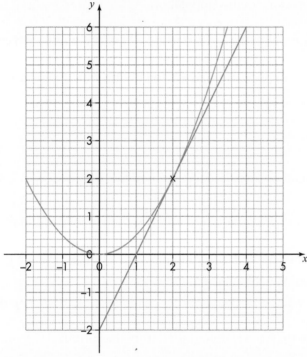

37 The headmaster of a new school offered his staff an annual pay increase of 5% for every year they stayed with the school.

a Mr Speed started teaching at the school on a salary of £28 000.

What salary will he be on after 3 years if he stays at the school?

b Miss Tuck started teaching at the school on a salary of £14 500.

How many years will it be until she is earning a salary of over £20 000?

38 A sycamore tree is 4 m tall. It grows at a rate of 8% per year. A conifer is 2 m tall. It grows at a rate of 15% per year.

How many years does it take before the conifer is taller than the sycamore?

39 **a** How long would it take to double your investment with an interest rate of 4% per annum?

b A ball bounces to $\frac{3}{5}$ of its previous height after each bounce. It is dropped from 10 m. How many times does it bounce before it bounces to just over 1 m above the ground?

c Write a similar problem but in a different context.

40 The exponential function is a function of the type $f(x) = a(b)^x$.

Many physical phenomena are represented by exponential growth and decay. One example is the population of bacteria.

a Assume a population in which each bacterium divides once per day.

Starting with a single bacterium at day 0, how many bacteria would you expect by day 6?

Explain your answer.

b Explain which values are constants and which are variables and why.

Explain what each represents in this population.

41 Use the function $F(x) = a(b)^x$, where a is the starting size of the population and b is the factor by which the function increases each time.

Explain what will happen to a population modelled on this function when:

   i $b < 1$

   ii $b = 1$

   iii $b > 1$

42 You can model disease mathematically, using the iterative formula:

$X_{n+1} = R^t x_0$

where R is the average number of people an infected person infects at time t since the start of an epidemic.

a Assume t is measured in days and the epidemic was started by a single carrier.

How many people would be infected after 10 days?

b Write a newspaper headline to engage readers with the story of this epidemic.

43 Atmospheric air pressure on Earth is determined from the formula:

$p = e^{-\left(\frac{h}{7}\right)}$

where p = atmospheric pressure (in bars), h = height or altitude above mean sea level (in km) and e is Euler's number.

Use an approximation to Euler's number of $e = 2.72$ and a spreadsheet to estimate the pressure at an altitude equivalent to the summit of Mount Kilimanjaro, 5895 metres.

44 **a** Show that $x = 1 + \dfrac{11}{x - 3}$ is a rearrangement of the equation $x^2 - 4x - 8 = 0$.

b Iteration is a way of solving equations. You would usually use iteration when you cannot solve the equation any other way.

Use the iterative formula $x_{n+1} = 1 + \dfrac{11}{x_n - 3}$ together with this graph of $x^2 - 4x - 8 = 0$

to obtain a root of the equation $x^2 - 4x - 8 = 0$ accurate to two decimal places.

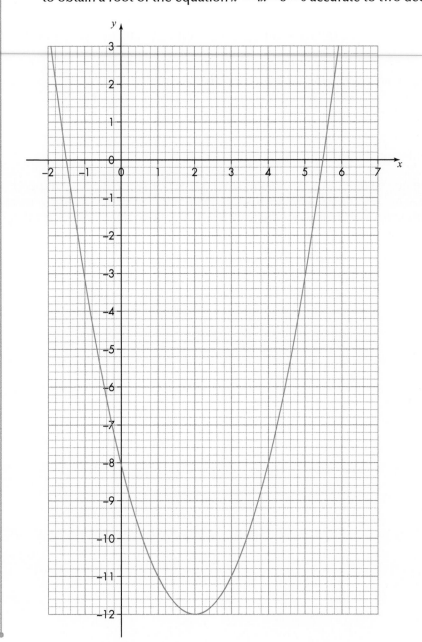

⑦Hints and tips

Question	Hint
5	Density = mass ÷ volume
7b	Make sure you read the question carefully. How much would everyone pay if Richard paid £2?
7c	Think about what you did in part **a** and work backwards.
17b	Think about what you did in part **a** and work backwards.
40b	Think about your starting value and your rate of increase.
43b	What is the newspaper trying to do, inform or sensationalise? Would it depend on whether you are writing for a broadsheet or a tabloid? You could state how many people might be infected after x days or how long it would take to infect your school, town, London, UK, the world.

Worked exemplars

1 A, B, C and D are four points on a number line.

A   B   C   D

AB : BC = 7 : 3

BC : CD = 2 : 5

Work out the ratio AC : CD.

Give your answer in its simplest form.

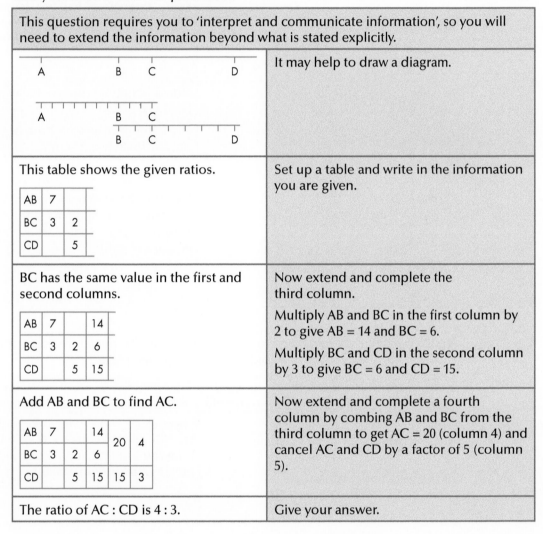

This question requires you to 'interpret and communicate information', so you will need to extend the information beyond what is stated explicitly.

It may help to draw a diagram.

This table shows the given ratios.

AB	7	
BC	3	2
CD		5

Set up a table and write in the information you are given.

BC has the same value in the first and second columns.

AB	7		14
BC	3	2	6
CD		5	15

Now extend and complete the third column.

Multiply AB and BC in the first column by 2 to give AB = 14 and BC = 6.

Multiply BC and CD in the second column by 3 to give BC = 6 and CD = 15.

Add AB and BC to find AC.

AB	7		14	20	4
BC	3	2	6		
CD		5	15	15	3

Now extend and complete a fourth column by combing AB and BC from the third column to get AC = 20 (column 4) and cancel AC and CD by a factor of 5 (column 5).

The ratio of AC : CD is 4 : 3.

Give your answer.

2 The mass of a solid, M, is directly proportional to the cube of its height, h.

When $h = 10$, $M = 4000$.

The surface area, A, of the solid is directly proportional to the square of the height, h.

When $h = 10$, $A = 50$.

Find the value of A when $M = 32\,000$.

This is a problem-solving question that requires you to translate a non-mathematical problem into a series of mathematical processes.	
$M = kh^3$ $4000 = k \times 10^3$ $\Rightarrow k = 4$ $M = 4h^3$	Set up the proportionality statement for the mass and height, then use the given information to find the value of k.
$A = ph^2$ $50 = p \times 10^2$ $\Rightarrow p = \dfrac{1}{2}$ $A = \dfrac{1}{2}h^2$	Now set up the proportionality equation for the area and the height. Use a different letter for the constant of proportionality constant.
$32\,000 = 4h^3$ $h^3 = 8000$ $\Rightarrow h = 20$ $A = \dfrac{1}{2} \times 20^2$ $= 200$	Use $M = 32\,000$ to work out the corresponding value of h, then substitute this into the proportionality equation for A.

3 A golden figure is melted down to create a million similar miniature figures that are all 3.5 cm tall. How tall was the golden figure in the first place?

This is a problem-solving question. You need to recognise that you know the ratio of volumes of similar shapes and can therefore work out what the ratio of lengths is.	
Ratio of volumes $= 1 : 1\,000\,000$ Ratio of lengths $= \sqrt[3]{1} : \sqrt[3]{1\,000\,000}$ $\qquad\qquad\quad = 1 : 100$ So height of original statue $\qquad\qquad = 3.5\text{ cm} \times 100$ $\qquad\qquad = 350\text{ cm}$	Show clearly how you work out the length ratio by finding cube roots. Then show how you use that ratio to calculate the height of the original statue.

(MR) 4 **a** Match each graph with the correct proportionality equation.

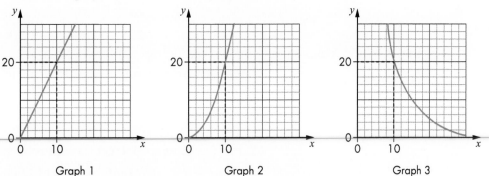

Graph 1   Graph 2   Graph 3

Equation A: $y = kx^2$ Equation B: $y = \frac{k}{x^2}$ Equation C: $y = kx$

(CM) **b** In each case work out the value of y when $x = 20$.

This is a mathematical reasoning question, so you need to demonstrate that you can apply your mathematical skills and knowledge to answer it.	
a Graph 1 matches equation C. Graph 2 matches equation A. Graph 3 matches equation B.	Start with the graph with which you are most confident. This is the linear graph. When you have matched that one, look at the quadratic. This will leave only one graph and equation remaining.
b Equation A $20 = k \times 10^2$ $\Rightarrow k = \frac{1}{5}$ $y = \frac{1}{5} \times x^2$ $y = \frac{1}{5} \times 20^2$ $= 80$	Set up the proportionality equation for each equation. Check that your graphs are correct in part **a** by substituting in values.
Equation B $20 = \frac{k}{10^2}$ $\Rightarrow k = 2000$ $y = \frac{2000}{x^2}$ $y = \frac{2000}{20^2}$ $= 5$	
Equation C $20 = k \times 10$ $\Rightarrow k = 2$ $y = 2x$ $y = 2 \times 20 = 40$	

5 Given that $y \propto x^2$ and that $y \propto \dfrac{1}{\sqrt{z}}$, which of these statements is true?

$$x \propto z \qquad x \propto \frac{1}{z} \qquad x \propto \frac{1}{\sqrt{z}} \qquad x \propto \frac{1}{\sqrt[4]{z}}$$

This question is about mathematical reasoning and is multi-choice.	
$y = kx^2$ and $y = \dfrac{K}{\sqrt{z}}$ $kx^2 = \dfrac{K}{\sqrt{z}}$	Set up the proportionality equations. Ignore the actual values of the proportionality constants as you do not need to work them out. Equate the two expressions for y.
$\sqrt{x^2} = \sqrt{\dfrac{K}{\sqrt{z}}} \Rightarrow x = \dfrac{\sqrt{K}}{\sqrt[4]{z}}$	Take the square root to get x as the subject. Because K just represents a constant, you can continue to use the same letter for the unknown value.
$x \propto \dfrac{1}{\sqrt[4]{z}}$	Choose the correct proportionality relationship.

4 Geometry and measures

1 a Is it possible to draw a triangle with:

   i one acute angle

   ii two acute angles

   iii one obtuse angle

   iv two obtuse angles?

  b Give an example of the three angles if it is possible. Explain why if it is impossible.

  c Explain why a triangle cannot have two parallel sides.

2 Is the following statement always, sometimes or never true? Justify your answer.

'When one rectangle has a larger perimeter than another one, it will also have a larger area.'

3 Right-angled triangles have half the area of the rectangle with the same base and height. Is this also true for non-right-angled triangles? Justify your answer.

4 Triangle A is drawn on a grid. Triangle A is rotated to form a new triangle B.

The coordinates of the vertices of B are (3, –1), (1, –4) and (3, –4).

Describe fully the rotation that maps triangle A onto triangle B.

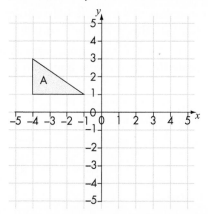

5 A council wants the outside of a block of flats repainted. The windows are uPVC, so do not need painting. A builder has been asked to provide a quote for painting the building.

The front and back of the building are 12 m high by 25 m wide. The sides of the building are 12 m high by 12 m wide.

There are 40 openings for windows and doors. Each measures 2 m by 1 m.

There are to be two coats of paint. Cans of paint cost £25 and contain 10 litres of paint.

1 litre of paint will cover 16 m².

The builder thinks it will take 2 weeks to do and he will need 3 painters.

Each painter costs £120 per day.

The scaffolding costs £500.

The builder adds 10% to all the costs to cover his time.

What should the builder charge the council, including 20% VAT?

6 This is a solid cube with side length 4 m. A cylinder with diameter 2.4 m is cut through from the front face to the back face.

   a What is the remaining surface area of the faces of the cube?

   b What is the volume of the shape?

   c The outside of the cube is going to be painted light blue.
The inside of the cylinder cut through the cube will be painted dark blue. 1 litre of paint covers 9 m².

     How much of each colour paint will be needed?

7 Shehab says: 'As long as I know the lengths of two sides of a triangle and the angle between them then I can draw the triangle.'

   Is Shehab correct? If not, explain why not.

8 You are asked to construct a triangle with sides 9 cm and 10 cm and an angle of 60°. Sketch all the possible triangles that you could construct from this description.

9 The locus of a point is described as 5 cm away from point A, and equidistant from both points A and B. Which of the following could be true? Explain your answer.

   a The locus is an arc.

   b The locus is just two points.

   c The locus is a straight line.

   d The locus is none of these.

10 **a** How does knowing the sum of the interior angles of a triangle help you to find the sum of the interior angles of a quadrilateral? Will this work for all quadrilaterals?

   b The angle at the vertex of a regular pentagon is 108°. Two diagonals are drawn to the same vertex to make three triangles. Calculate the sizes of the angles in each triangle.

11 One of the lines of symmetry of a regular polygon goes through two vertices of the polygon. Explain why the polygon must have an even number of sides.

12 Sketch a shape to show that:

   a a trapezium might not be a parallelogram

   b a trapezium might not have a line of symmetry

   c every parallelogram is also a trapezium.

13 Is this statement sometimes, always or never true? Explain your answer.

   'The sum of the exterior angles of a polygon is 360°.'

14 The angles in a triangle are in the ratio 6 : 5 : 7.

   Work out the sizes of the three angles.

15 This is part of a design for a new company logo. The small circle has a radius of 1.5 cm and the large circle has a diameter of 12 cm.

   Use this information to work out a reasonable estimate for the shaded area. Make sure you explain any assumptions you make.

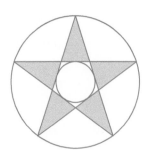

16 **a** Explain why equilateral triangles, squares and regular hexagons will tessellate on their own, but other regular polygons will not.

    b Explain why squares and regular octagons will tessellate.

17 What is the minimum information you need about a triangle to be able to calculate all three sides and all three angles?

18 Are these statements true or false? Justify your answers.

    a Every rhombus is a parallelogram, and a rhombus with right angles is a square.

    b A rhombus must be a parallelogram but a parallelogram is not necessarily a rhombus.

    c A trapezium cannot have three acute angles.

    d A quadrilateral can have three acute angles.

19 How do you decide whether you need to use a trigonometric relationship (sine, cosine, tangent, sine rule or cosine rule) or Pythagoras' theorem to solve a triangle problem?

20 **a** **i** Make up a simple reflection.

      ii Make up a more complicated reflection. What makes it harder than your first one?

    b **i** Make up a simple rotation.

      ii Make up a more complicated rotation. What makes it harder than your first one?

21 **a** What changes when you enlarge a shape? What stays the same?

    b What information do you need to complete a given enlargement?

    c How do you find the centre of enlargement and the scale factor of an enlargement?

    d **i** Explain how the position of the centre of enlargement (e.g. inside, on a vertex, on a side, or outside the original shape) affects the image.

      ii Explain what happens to the image when the scale factor is between 0 and 1.

22 **a** How can you tell if a shape has been reflected or translated?

    b Describe how the image produced by rotating a rectangle about its centre looks different from the image produced by rotating it about one of its vertices.

23 The cross-section of a length of skirting board is in the shape of a rectangle, with a quadrant (quarter circle) on the top. The skirting board is 1.5 cm thick and 6.5 cm high. Lengths totalling 120 m are ordered. What is the volume of wood in the order?

24 Prove that the area of a segment of a circle, where the angle at the centre is equal to θ, is $r^2 \left(\frac{\theta}{360} \pi - \frac{1}{2} \sin \theta \right)$.

25 Two circles of radius 6 cm overlap. The centres of the two circles are 10 cm apart. Estimate the area of the overlap.

26 Starting from this 2D representation of a 3D shape:

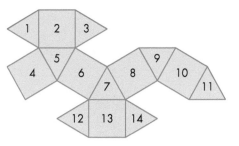

  a How many faces will the 3D shape have? How do you know?

  b Which face will be opposite face numbered 2 in the 3D shape? How do you know?

  c How would you draw the plan and elevation for the 3D shape you could make from this net?

Real-life problems are often quite complicated and use a lot of words. It is important you are able to decode the words so you can work out the mathematics. Question 27 is an example of this. Remember to think carefully what the words mean and what mathematics you will need to use.

27 A cycle has wheels of diameter 68 cm.

How many complete rotations does each wheel make for every 10 km travelled?

28 You walk due north for 5 km, then due east for 3 km. What is the shortest distance you could be from your starting point? Justify your answer.

29 A surveyor wishes to measure the height of a chimney. Measuring the angle of elevation, she finds that the angle increases from 28° to 37° after she walks 30 m towards the chimney. What is the height of the chimney?

30 Ship S and two lighthouses A and B are shown in the diagram below. A is due west of B and the two lighthouses are 15 km apart.

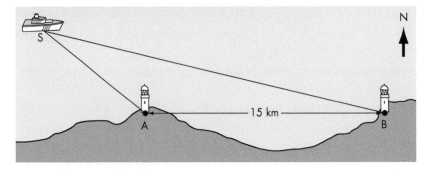

The bearing of the ship from lighthouse A is 330° and the bearing of the ship from lighthouse B is 290°.

Calculate the distance of the ship from lighthouse A.

31 **a** 'You can build a solid cuboid using a given number of identical interlocking cubes.' Is this statement always, sometimes or never true?

  b If it is sometimes true, when is it true and when is it false? For what numbers can you only make one cuboid?

  c For what numbers can you make several different cuboids?

32 Is it possible to slice a cube so that the cross-section is:

a a rectangle

b a triangle

c a pentagon

d a hexagon?

It is very important to be able to recognise types of questions. This will save you time and give you confidence. One common type of question you will be asked is packing questions.

33 A cuboid has surface area of 96 cm². When you double the length of its sides you double its area and multiply its volume by 3. Is this true or false? Justify your answer.

34 A picture is hanging on a string secured to two points at the side of the frame.

The string is initially 45 cm long.

When the picture is hung the string stretches as shown.

By how much does the string stretch?

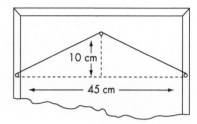

35 Three towns, A, B and C, are joined by two roads, as in the diagram. The council wants to build a road that runs directly from A to C. How much distance will the new road save? Give your answer to one decimal place.

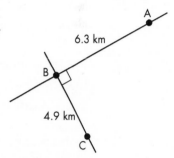

36 Is it possible to have a triangle with the angles and lengths shown? Explain your answer.

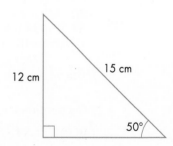

37 Use what you know about sine 45° to prove that this is an isosceles triangle.

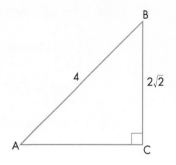

38 Use Pythagoras' theorem to work out the length of AB in this triangle. Leave your answer in surd form.

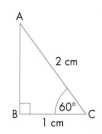

39 Which of these ratios is the odd one out and why?

cos 68°   cos 112°   cos 248°   cos 338°

40 **a** Solve:

    i $\sin x + 1 = 2$ for $0° < x < 360°$

    ii $2 + 3\cos x = 1$ for $0° < x < 360°$

  b Find two values of x between 0° and 360° such that $\sin x = \cos 320°$.

41 The sine of x is $\dfrac{\sqrt{8}}{4}$.

The cosine of x is $\dfrac{3}{\sqrt{18}}$.

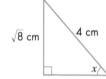

  a Work out the tangent of x.

  b Work out the size of x.

42 In a right-angled triangle the two sides that are at right angles to each other are $\sqrt{6}$ cm and $\sqrt{10}$ cm long.

Show clearly that $(\sin x)^2 + (\cos x)^2 = 1$.

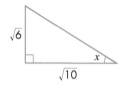

43 In the triangle ABC, AC = 3 cm, BC = 4 cm and angle ACB = 60°. Work out the length of AB, giving your answer in surd form.

44 A group of hikers walk between three points A, B and C, using vectors with distances in kilometres.

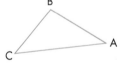

The vector $\overrightarrow{AB}$ is $\begin{pmatrix} -4 \\ 3 \end{pmatrix}$ and the vector $\overrightarrow{BC}$ is $\begin{pmatrix} -2 \\ -5 \end{pmatrix}$.

On centimetre-squared paper, draw a diagram to show the walk, using a scale of 1 cm to represent 1 km. Work out the vector $\overrightarrow{CA}$.

45 Joel says that when the translation from a point X to a point Y is described by the vector $\begin{pmatrix} -3 \\ 2 \end{pmatrix}$, the translation from the point Y to the point X is described by the vector $\begin{pmatrix} 2 \\ -3 \end{pmatrix}$.

Is Joel correct? Explain how you decide.

46 Using a straight edge and a pair of compasses only, construct:

  a an angle of 15 degrees

  b an angle of 75 degrees.

47 When you construct all the angle bisectors in a triangle, they meet at a point.

Explain why you can draw a circle inside the triangle with this point as the centre, and why this circle will just touch each side of the triangle.

48 This construction is the nine-point circle of a triangle.

a Draw a triangle ABC.

b Construct the midpoints of the three sides and call these L for AB, M for BC and N for AC.

c Construct the perpendiculars to the opposite sides from the vertices A, B and C. Call the point where they intersect O. Label the feet of the perpendiculars D, E and F on AB, BC and AC, respectively.

d Construct the midpoints of AO, BO and CO. Label these X, Y and Z, respectively.

e Bisect the line segments LM, LN and MN. Call the point where they intersect P.

f Can you think of a way to check all nine points you have drawn?

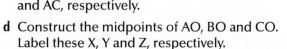

49 In the diagram, AB and CD are parallel with AB = CD. The lines AC and BD intersect at X.

Prove that triangle ABX and triangle CDX are congruent.

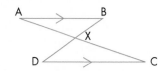

50 Helen says that these two triangles are congruent because the three angles are the same.

Explain why she is wrong.

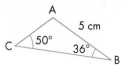

51 **a** Imagine a square. Join the midpoints of each of the adjacent sides. What is the inscribed shape? How do you know? Justify your answer.

b What is the area of the inscribed shape compared to the original shape? Justify your answer.

52 ABCDEFGH is a regular octagon. HJFKLM is a regular hexagon.

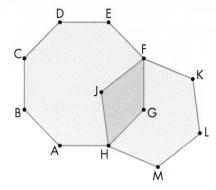

Work out the size of angle JFG. Justify your answer.

53 ABC is a triangle. P is a point on BC such that angle
APC = angle BAC. The sides of ABC are a, b and c. AP = p
and PC = r.

Prove that: $r = \dfrac{bc}{a}$

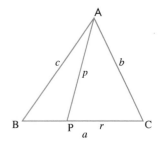

54 ACB and ADB are right-angled triangles. The lengths are marked in the diagram.

Use Pythagoras' theorem on both triangles ACB and ADB to prove that: $xt = sy$.

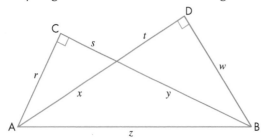

55 **a** Write an explanation to show how you can use Pythagoras' theorem to tell
whether an angle in a triangle is equal to, greater than or less than 90 degrees.

b What is the same/different about a right-angled triangle with sides 5 cm, 12 cm and an
unknown hypotenuse, and a right-angled triangle with sides 5 cm, 12 cm and an
unknown shorter side?

56 The midpoints of the edges of a square are each joined to a vertex to create a smaller
square, shown shaded in the diagram. Explain why the shaded square has an area
one-fifth of the area of the larger square.

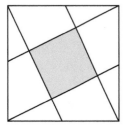

57 Which of these statements are true? Explain your reasoning.

a Any two right-angled triangles will be similar.

b When you enlarge a shape you get two similar shapes.

c All circles are similar.

58 Write a justification that will convince your teacher that:

a any two regular polygons with the same number of sides are similar

b alternate angles are equal (using congruent triangles).

59 Copy the diagram onto squared paper.

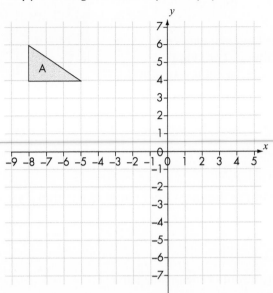

a Triangle A is translated by the vector $\begin{pmatrix} 9 \\ -3 \end{pmatrix}$ to give triangle B. Draw triangle B.

b Triangle B is then enlarged by a scale factor –2 about the origin to give triangle C. Draw triangle C.

c Describe fully the single transformation that maps triangle C onto triangle A.

60 Triangle B is an enlargement of triangle A.

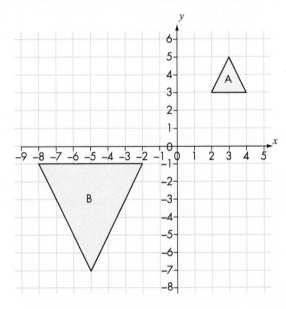

Which of the following describes the enlargement?

an enlargement of scale factor –2 about (0, 0)

an enlargement of scale factor –3 about (0, 0)

an enlargement of scale factor –3 about (1, 2)

an enlargement of scale factor $-\frac{1}{3}$ about (1, 2)

Explain how you decide.

61 When letters are taken to a sorting office they are checked to see if there is a stamp in the top right-hand corner.

Each letter is put through a checking machine and the top right-hand corner is scanned. The stamp will be detected only if the letter is the right way round.

If no stamp is detected, the letter is rotated automatically and put through the machine again.

Two different rotations are used:

180° rotation about the horizontal line H

180° rotation about the vertical line V

These rotations will leave the letter in the same orientation, but with the stamp in the other corner.

Here is the procedure used:

Scan the letter.

If no stamp is detected, rotate about H and scan again.

If no stamp is detected, rotate about V and scan again.

If no stamp is detected, rotate about H and scan again.

If no stamp is detected then the letter is rejected as unstamped.

a Show that if this procedure is followed then any letter that is correctly stamped will be detected, whichever way the letter is initially fed into the machine.

b If someone accidentally put the stamp on the top left-hand corner of the letter, would the machine detect it?

c Suppose that a new regulation requires all letters to be square. Explain why there are now eight corners to be checked for a stamp.

d With a square letter you could use rotations about either of the diagonals as well as the horizontal and vertical lines. Show how it is still possible to use just two rotations (repeated if necessary) to check all eight corners.

62 Prove that the perpendicular from the centre of a circle to a chord bisects the chord.

63 Prove that angles in same segment of a circle are equal.

64 On the diagram, O is the centre of the circle.
Angle BAC = x and angle CBO = y.

Prove that $y = x - 90°$, giving reasons in your working.

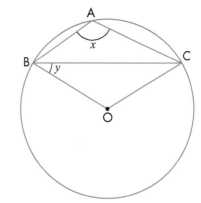

65 ABCD is a cyclic quadrilateral.
Work out the values of x and y.

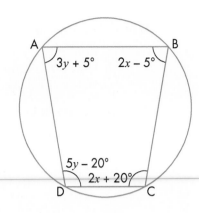

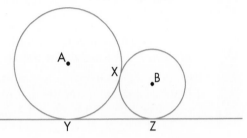

66 The diagram shows two circles touching at X.
The circles have a common tangent at Y and Z.
The circle with centre A has a radius of 6 cm.
The circle with centre B has a radius of 3 cm.
Calculate the length YZ.

67 Prove that the angles subtended by a chord at the circumference of a circle are equal.

68 PQRS is a cyclic quadrilateral. PR and QS meet at T.
   a Work out the value of x.
   b Show that the angles of the quadrilateral and angle STP form a number sequence.

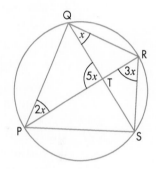

69 The diagram shows a hollow closed glass container containing some coloured water.
Show that if the cuboid is placed on a different face the depth of water stays a constant proportion of the vertical height.

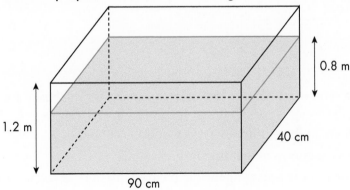

70 **a** When you know the height and volume of a right prism, what else do you know? What don't you know?

b How many different square-based right prisms have a height of 10 cm and a volume of 160 cm³? Explain your answer.

A formula that just includes length on its own is a formula for **length**.

A formula with a length multiplied by another length is for an **area**.

A formula with length × length × length will be for **volume**.

Which formula is for the volume of a cube? Which formula is for the surface area of a cuboid?

a^3

$2ab + 2bc + 2ac$

The first equation involves length × length × length. It fits the definition for volume and therefore is the formula for the volume of a cube.

The second equation involves three different quantities added together. Each of these quantities involves multiplying one length by another. Therefore this must be the formula for the surface area of the cuboid.

71 **a** Why is it easy to distinguish between the formulae for the circumference and area of a circle?

b How would you help someone to distinguish between the formula for the surface area of a cube and the formula for the volume of a cube?

c Here are two formulae:

$$\frac{1}{3}b^2h \qquad 4\pi r^2$$

One is for the surface area of a sphere and the other is for the volume of a square-based pyramid. Which formula is which? Use Example 1 to help you explain.

72 The length of string needed to tie up a package in the shape of a cuboid is given by the formula:

$S = 2L + 2W + 4H + 20$

where L is the length of the package, W is its width and H is its height. All lengths are in centimetres.

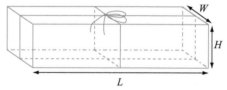

Masood wants to send eight identical boxes in one package. The boxes are cubes with side length 15 cm.

He can arrange them in three different ways to make a cuboid.

Two of these ways are shown in the diagram.

Which of the three ways of arranging the eight cubes in a cuboid will use the least amount of string?

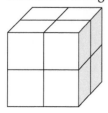
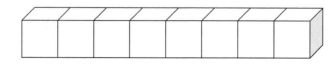

73 A sweet manufacturer wants a new package for an assortment of sweets. The package must have a volume of 1000 cm³ to hold the sweets. The chosen design will be a prism. The length has been specified as 20 cm.

The cross-section of the package will be a square, an equilateral triangle or a circle.

You have been asked to investigate the amount of packaging material needed for each design, because this will affect the cost of manufacture. Calculate the surface area of each of the three designs.

Comment on how much difference there is between the three surface areas and how this could affect production costs.

This formula could be useful:

area of an equilateral triangle of side a is $\frac{1}{4}\sqrt{3}a^2$.

74 Harry is travelling on a road that goes directly from X to Y. The road is closed between A and B because of flooding. Harry has to make a detour through C.

Calculate how much further Harry has to travel by making the detour.

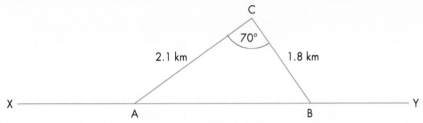

75 A TV mast XY is 3 km due west of village A. Village B is 2 km due south of village A. The angle of elevation from ground level at B to the top of the mast is 6°. Show how you can use this information to calculate the height of the mast, in metres.

76 A tetrahedron VPQR stands on a prism FGHPQR. The cross-section PQR is an equilateral triangle of side 8 cm. VP = VQ = VR = 10 cm. PF = QG = RH = 15 cm. M is the midpoint of QR.

a Use triangle PQR to find the length of PM.

b Use triangle VQR to find the length of VM.

c Find the size of angle VPM.

d Find the height of V above the base FGH. Give your answer to an appropriate degree of accuracy.

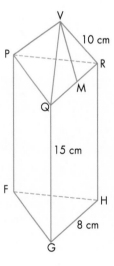

77 XABCD is a right pyramid on a rectangular base.

Ben is working out the angle between the edge XA and the base ABCD.

This is his working:

By Pythagoras' theorem

$$AC^2 = 12^2 + 8^2 = 208$$

$$\text{so } AC = \sqrt{208}$$

Let angle XAC = x

$$\cos x = \frac{\sqrt{208}}{10}$$

so $x = \cos^{-1}\dfrac{\sqrt{208}}{10}$

Ben gets an error message on his calculator when he tries to work this out.

Explain where Ben has made the error and write out a correct solution to find the value of x.

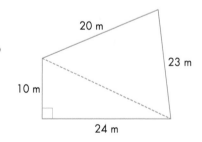

78 A vertical tower CD sits on horizontal ground ABC. The angle of elevation of D from A is 32°. The angle of elevation of D from B is 58°. AB is 28 m. Calculate the height of the tower.

79 Zainab leaves a hostel H and walks 8.6 km on a bearing of 128° to a checkpoint C. She then walks 4.7 km on a bearing of 066° to the next hostel B. The next day she returns to the original hostel to pick up her luggage. She walks directly back to hostel H. How far does she need to walk and on what bearing?

80 In the triangle ABC, angle B is obtuse, ∠BAC = 32°, AC = 10 cm, BC = 6 cm. Calculate the area of the triangle ABC.

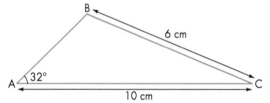

81 The diagram shows a plan of a farmer's orchard.

a Calculate the area of the orchard. Give your answer to an appropriate degree of accuracy.

b For each 5 m² the farmer will plant a tree. How many trees can the farmer plant in the orchard?

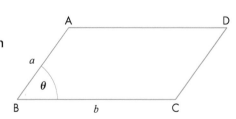

82 ABCD is a parallelogram. AB = a and BC = b.
Angle ABC = θ.

Prove that the area, A, of the parallelogram is given by the formula:

$A = ab \sin \theta.$

83 ABCD is a quadrilateral.

Work out the area of the quadrilateral. Give your answer to an appropriate degree of accuracy.

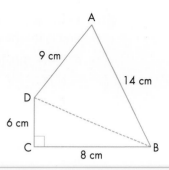

84 A formula for the area of a triangle with sides a, b and c is:

$$\text{area} = \sqrt{s(s-a)(s-b)(s-c)}$$

where $s = \frac{1}{2}(a + b + c)$

a Use this formula to find the area of a right-angled triangle with integer sides and show that it gives the correct answer.

b Use the formula to find the area of an equilateral triangle. Find the area of this triangle by a different method and show that the formula gives the correct answer.

c A piece of land is roughly triangular in shape. The sides of the triangle are 18 metres, 22 metres and 24 metres. Work out the area of the piece of land.

d A field is a quadrilateral in shape. You have been asked to find the area of the field. You have a long measuring tape to help you. What measurements would you take and how would you use them to find the area of the field?

85 Write down a series of translations that will take you from the Start/finish, around the shaded square without touching it, and back to the Start/finish. Make as few translations as possible.

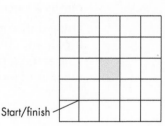

86 $\overrightarrow{OA} = \mathbf{a}$ and $\overrightarrow{OB} = \mathbf{b}$

a Express each of these vectors in terms of $\mathbf{a}$ and $\mathbf{b}$. Give your answers in their simplest form.

   i $\overrightarrow{AB}$

   ii $\overrightarrow{AD}$

   iii $\overrightarrow{EC}$

   iv $\overrightarrow{FB}$

b Write down two facts about the lines EC and FB.

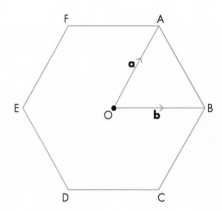

87 A, B and C are three points with $\overrightarrow{AB} = 6\mathbf{a} + 4\mathbf{b}$ and $\overrightarrow{AC} = 9\mathbf{a} + 6\mathbf{b}$.

a Write down a fact about the points A, B and C. Give a reason for your answer.

b Write down the ratio of the lengths AB : BC in its simplest form.

88 OABC is a quadrilateral. $\overrightarrow{OA}$ = **a**, $\overrightarrow{OB}$ = **b**, $\overrightarrow{OC}$ = **c**. M, N, Q and P are the midpoints of OA, OB, CB and AC, respectively.

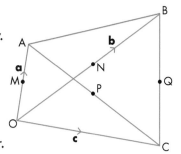

a Work out, in terms of **a**, **b** and **c**, the vectors:

  i $\overrightarrow{BC}$

  ii $\overrightarrow{NQ}$

  iii $\overrightarrow{MP}$

b What type of quadrilateral is MNQP? Explain your answer.

89 On the diagram, $\overrightarrow{OA}$ = **a**, $\overrightarrow{OB}$ = **b** and $\overrightarrow{OC}$ = 3**b** − 2**a**.

Prove that ABC is a straight line.

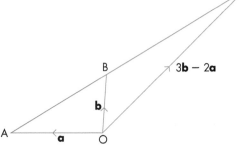

90 A knight is on the white square near the bottom left-hand corner of a chess board. The two possible moves it can make are shown by the vectors **a** and **b**.

Using only combinations of these two types of move, how many different ways can the knight reach the king at the top of the chess board in the minimum number of moves?

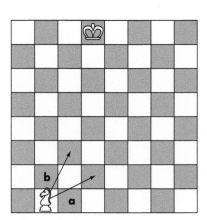

⑦Hints and tips

Question	Hint
1	Use examples and diagrams to help your explanation.
10	Use a diagram and/or example to help.
21	A diagram might help your justification.
25	You may want to refer to your answer to Q24.
42	Draw a diagram to help you.
51	
53	Show that triangles ABC and APC are similar.
62	Check this list of circle theorems.

The angle at the centre of a circle is twice the angle at the circumference when they are both subtended by the same arc.

Every angle subtended at the circumference of a semicircle by the diameter is a right angle.

Angles subtended at the circumference in the same segment of a circle are equal.

Opposite angles in a cyclic quadrilateral sum to 180°.

The angle between a tangent to a circle and its radius is 90°.

Tangents to a circle from an external point to the points of contact are equal in length.

The line joining an external point to the centre of the circle bisects the angle between the tangents.

The perpendicular from the centre to a chord bisects the chord.

Alternate segment theorem. |
63	You will need to use at least one of the other circle theorems.
72	This question is a packing question. Think about how many of one 3D shape you need to fill another 3D shape.
78, 79	Draw a diagram first to help you.

Worked exemplars

1 ABC is a triangle. D is a point on AB such that BC = BD.

a Work out the value of *x*.

b Work out the value of *y*.

c Is it true that AD = DC? Give a reason for your answer.

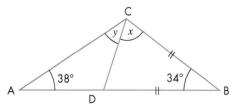

This is a question where you are required to communicate mathematically. You will need to show clearly how you have found the indicated angles and clearly explain your final response to part **c**.	

a Triangle BCD is isosceles, so angle BDC is equal to *x*.
 	
  Angles in a triangle = 180°
 	
  Therefore, $2x + 34° = 180°$
                    $2x = 146°$
                      $x = 73°$
	First, set up an equation in *x* from the knowledge that angles in a triangle add up to 180° and then solve.

b Method 1
  Angle ADC = 180° − 73°
               = 107° Angles on a line
  $y + 38° + 107° = 180°$ Angles in a triangle
       $y + 145° = 180°$
                $y = 35°$
Method 2
  Angle ACB = 180° − (38° + 34°)
               = 108° Angles in a triangle
  $y + x$ = Angle ACB
       = 108°
    $y = 108° − 73°$
       = 35°
	To find the value of *y*, you need to show how you are using the given angles and the found value of *x*. You should show the mathematical reasoning used at each stage. There are two ways of calculating *y* here. Both are acceptable.

c No, since triangle ACD is not an isosceles triangle, no two sides of the triangle are equal.	Clearly state your explanation about ACD not being isosceles. The answer 'No', alone, is not sufficient.

2 Hari has a cylindrical glass and another glass in the shape of a cone connected to a stem.

They both hold the same amount of liquid.

The cylindrical glass has a diameter of 5 cm and is 6 cm high.

The other glass has an opening of diameter of 6.4 cm and a stem height of 3 cm.

How high is this glass?

This is a problem-solving question. You need to recognise that you must find the volume of one glass and then set up an equation to help find the height of the other glass.	
The cylindrical glass has a base radius of 2.5 cm. The volume is $\pi \times 2.5^2 \times 6 = 37.5\pi$ cm³.	First, find the volume of the cylindrical glass.
If the height of the cone in the second glass is h cm then: $\frac{1}{3} \times \pi \times 3.2^2 \times h = 37.5\pi$ $h = \frac{37.5 \times \pi \times 3}{\pi \times 3.2^2}$ $\quad = 10.9863$ cm.	Then set up an equation and solve it to find the height of the other glass. Note that you don't need to complete the calculations fully – you can leave them in terms of π, as π will cancel out.
The height of the glass is $11 + 3$ cm $= 14$ cm.	You need to round the final answer appropriately.

 3 A clock is designed to have a circular face on a triangular surround.
The triangle is equilateral.

The face extends to the edge of the triangle.

The diameter of the clock face is 18 cm.

Show that the perimeter of the triangle is 94 cm.

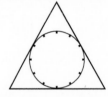

This is a communicating mathematics question where you have to construct a chain of reasoning to achieve a given result.	
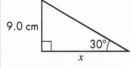 The radius is 9 cm. The angle is 30° because it is half the angle of an equilateral triangle. $\tan 30° = \frac{9}{x}$ $\Rightarrow x = \frac{9}{\tan 30°}$	You have to find the strategy of getting to the given result of 94 cm, clearly showing your method at each stage. You need to show a correct trigonometry ratio that can be used to calculate x, half the length of the side of the triangle. Use the lengths you are given, and what you can deduce, to draw a triangle.
Perimeter of triangle $= 6 \times \dfrac{9}{\tan 30°}$ $= 93.53\ldots$ cm $= 94$ cm (2 sf)	Show the correct value of 93.53 and how you rounded to 2 sf in order to get the given solution of 94 cm.

4 A camping gas container is in the shape of a cylinder with a hemispherical top. The dimensions of the container are shown in the diagram.

A new design for the container increases the surface area by 15%, keeping the new container mathematically similar to the old one.

It is suggested that this new design makes the container contain 1000 cm³. Is this claim correct?

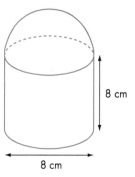

8 cm

8 cm

This is a communicating mathematics question where you are assessing the validity of a statement.

Old area : new area = 100% : 115% = 1 : 1.15 Length ratio = $\sqrt{1} : \sqrt{1.15}$ = 1 : 1.072 380 5 Volume ratio = $1^3 : 1.072\,380\,5^3$ = 1 : 1.233 237 6 Volume of container = $\pi r^2 h + \frac{1}{2} \times \frac{4}{3} \pi r^3$ Volume of old container = $\pi \times 16 \times 8 + \frac{4}{6} \times \pi \times 64$ = 536.165 15 cm³ Volume of new container = 536.165 15 × 1.233 237 6 = 661 cm³ (3 sf) This is much less than 1000 cm³. The claim would only be correct if the number was rounded to the nearest thousand.	You need to recognise that you have to use the area scale factor to get the length scale factor in order to work out the volume scale factor. You could calculate the new dimensions of the larger container and then calculate the volume, but the numbers would be awkward. It is simplest to work out the volume of the old container and then increase that volume by the found scale factor as here. You then have to evaluate the claim made. It is incorrect but it is good to recognise what could have made the claim correct.

 5 In the diagram, XY is a tangent to the circle at A.

BCY is a straight line.

Work out the size of ∠ABC.

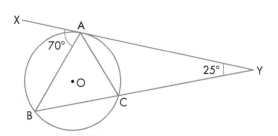

This is a problem-solving question where you need to plan your strategy to find the size of this angle.

∠ACB = 70°	Angle in alternate segment	You need to combine your knowledge of angles in a triangle with angles in a circle to identify the unknown angles in the diagram until you are able to calculate the angle ABC.
∠ACY = 110°	Angles on a line = 180°	
∠CAY = 45°	Angles in a triangle sum to 180°	
so ∠ABC = 45°	Angle in alternate segment	You need to show your reasoning at each stage: it is not sufficient simply to state the answer with no indication of how you found it.

6 The diagram shows a cuboid ABCDEFGH.

Ben said the size of angle AGE is 23°.

Evaluate Ben's statement.

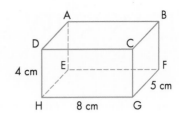

This is an evaluating question where you check the statement to see if it is correct. This means you need to calculate the given angle in order to check the validity of the statement.

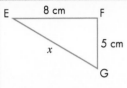 $x^2 = 8^2 + 5^2$ $\quad = 89$ $\Rightarrow x = \sqrt{89}$ $\quad = 9.434$ cm	Identify a triangle from the given diagram with the required angle: AGE. You need to find length EG in order to have sufficient data in the triangle to work out the angle. Draw the right-angled triangle EFG and work out the length of EG, using Pythagoras' theorem.
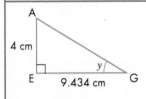 $\tan y = \dfrac{O}{A}$ $\quad = \dfrac{4}{9.434}$ $\quad = 0.4240$ $\Rightarrow y = \tan^{-1} 0.420$ $\quad = 23.0°$ (3 sf)	Draw the right-angled triangle AGE, label the required angle y and calculate its value.
So Ben is correct.	Once you have found the angle, say explicitly whether or not the statement is correct.

5 Probability

1 Are the following statements true or false? Justify your answer.

  a Experimental probability is more reliable than theoretical probability.

  b Experimental probability gets closer to the true probability as more trials are carried out.

  c Relative frequency finds the true probability.

2 Andrew made a six-sided spinner.

He tested it to see if it was fair. He spun the spinner 240 times and recorded the results in a table.

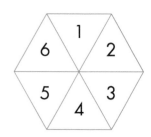

Number spinner lands on	1	2	3	4	5	6
Frequency	43	38	41	41	42	44

Do you think the spinner is fair? Give reasons for your answer.

3 Joy, Vicky and Max play cards together every Sunday night.

Joy is always the favourite to win, with a probability of 0.65.

There were 52 Sundays in the year and Vicky won 10 times.

How many times would you expect Max to have won in the year?

4 Two of these six people are to be chosen for a job.

Anna, Ben, Chloe, Clara, Ciaran, Daniel

  a List all of the possible pairs (there are 15 altogether).

  b What is the probability that the pair of people chosen will:

    i both be female

    ii both be male

    iii both have the same initial

    iv have different initials?

  c Which of these pairs of events are mutually exclusive?

    i Picking two women and picking two men.

    ii Picking two people of the same sex and picking two people of opposite sex.

    iii Picking two people with the same initial and picking two men.

    iv Picking two people with the same initial and picking two women.

  d Which pair of mutually exclusive events in part **c** is also exhaustive?

5 **a** Explain the key features of mutually exclusive and independent events when they are shown on a probability tree diagram.

  b Explain why the probabilities on each set of branches have to sum to 1.

  c How can you tell from a completed probability tree diagram whether the question specified with or without replacement? Use an example to help your explanation.

  d What strategies do you use to check the probabilities on your probability tree diagram are correct?

6 The weather forecast says the probability of rain on Thursday is 25% and is 48% on Saturday. What is the probability that it will rain on just one of the days?

7 Shehab walked into his local supermarket and saw a competition.

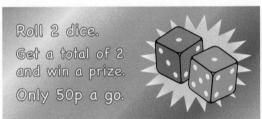

Roll 2 dice.
Get a total of 2
and win a prize.
Only 50p a go.

  a What is the probability of winning a £20 note?

  b How many goes should he have to expect to win at least once?

  c If he has 100 goes, how many times would he expect to win?

8 A fair spinner has three equal sections. The spinner is spun twice.

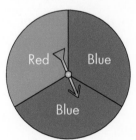

  a Show that the probability of scoring two blues is the same as for scoring one red and one blue.

  b This spinner has four equal sections. Is the probability of scoring one yellow and one green the same as for scoring two greens? Make sure you justify your answer.

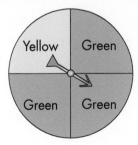

  c Design a spinner with equal sections where the probability of scoring two greens is the same as for scoring one green and one yellow.

9 Thomas takes a driving test which is in two parts.

The first part is theoretical. He has a 0.4 chance of passing this.

The second is practical. He has a 0.5 chance of passing this.

a Draw a probability tree diagram covering passing or failing the two parts of the test.

b What is the probability that he passes both parts?

10 In the game 'Rushdown', you are dealt two cards from a normal pack of cards. If you are dealt any two of the numbers 6, 7 and 8, you have been dealt a 'Tango'.

What is the probability of being dealt a 'Tango'? Give your answer to 3 decimal places.

11 The diagram shows a board for a game.

Joseph begins with his counter on Start. He rolls a fair dice.

He moves his counter one square to the right when the dice shows a 1, 2, 3 or 4. Otherwise he moves his counter one square to the left.

Show that landing on the square with 1 in it is twice as likely as landing on the square with −1 in it.

12 I throw a coin and roll a normal six-sided dice. The probability of scoring a 5 and a head is $\frac{1}{12}$. This is not $\frac{1}{2} + \frac{1}{6}$. Why not?

13 The probability that Nora fails her driving theory test on the first attempt is 0.1.

The probability that she passes her practical test on the first attempt is 0.6.

Complete a probability tree diagram based on this information. Use your diagram to find the probability that she passes both tests on the first attempt.

14 Anne regularly goes to London by train.

The probability of the train arriving in London late is 0.08.

The probability of the train being early is 0.02.

The probability of it raining in London is 0.3.

What is the probability of:

a Anne getting to London on time and it not raining

b Anne travelling to London three days in a row and it raining every day

c Anne travelling to London five days in a row and not being late at all?

15 In a test there are two questions. Problem one is solved by 50% of the students. Problem two is solved by 80% of students. Every student solves at least one of the problems. 12 students solve both problems.

How many students took the test? Use a Venn diagram and the equation to help you work this out.

$x = 0.8x - 0.5x - 12$

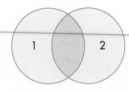

16 At a party 70% of people had brown hair and 30% of people had both brown hair and wore glasses. Thirty people wore glasses.

a How many people were at the party?

b Make up a problem like this of your own. Write a mark scheme for your question showing how you would use it.

17 I roll a normal six-sided dice four times and add the four numbers I score. Explain the difficulty in drawing a sample space to show all the possible events.

18 a Write your own problem that can be solved by adding probabilities.

b Write your own problem that can be solved by multiplying probabilities.

c Now write a problem that involves both adding and multiplying probabilities.

19 A social club runs a lottery, using the numbers 1 to 10. Three numbers are chosen each week and are not replaced.

a Margaret chooses the numbers, 6, 7 and 8.

What is the probability she wins the lottery?

b Richard says Margaret is never going to win with three consecutive numbers. He chooses the ages of his nieces and nephews which are 2, 5 and 7.

He says: 'I have a better chance than Margaret.'

Is he correct? Explain your answer.

c There are 45 members in the social club. Every week they all choose three numbers and pay 20p. If anyone's three numbers are selected, that person wins £5. There are no rollovers. Throughout the year, the lottery is run 50 times.

How much profit would the social club expect to make in one year?

20 I have a drawer containing black, blue and brown socks. The socks are not paired.

Explain how drawing a probability tree diagram can help me find the probability of picking at random two socks of the same colour.

21 Paul is playing a card game and is dealt three cards, all aces. He thinks the chance of him being dealt another ace is $\frac{1}{52}$. Explain why he is wrong.

22 In a pack of cards, the aces, the kings, the queens and the jacks are all called picture cards. What is the probability of being dealt four picture cards in a row from a normal pack of cards?

23 A bag of jelly babies contains only yellow, green and orange jelly babies, all the same size. Trisha is asked to find the probability of taking out two jelly babies of the same colour.

Explain to Trisha how you would do this. Explain carefully the point where she is most likely to go wrong.

24 Five friends each choose a random number in the range 1 to 9. What is the probability that any of them choose the same number?

25 There are 25 people at a party.

  a What is the chance that any two of them have their birthday on the same day?

  b Alison says: 'There are 25 people, and 365 days, so $\frac{25}{365}$ sounds about right, and $\frac{25}{365} = 0.07$.' Explain why she is not correct.

  c What assumption are you making and why might this not be true in reality?

26 There are 10 people in a room. What is the chance that the first names of any two of them start with the same letter of the alphabet?

  a What assumption would you need to make to be able to start this question?

  b Based on this assumption, how many people would you need in the room to make the theoretical probability more than 50% that the first names of any two of them start with the same letter of the alphabet?

⑦Hints and tips

Question	Hint
10	Draw the sample space for this event.
25	Assume 365 days in a year and the probability they do is 1 – (the probability that they don't).
26	Think about which letters the names of other students start with. Are some more common than others?

Worked exemplars

 1 In a raffle 400 tickets have been sold. There is only one prize.

Mr Raza buys 5 tickets for himself and sells another 40.

Mrs Raza buys 10 tickets for herself and sells another 50.

Mrs Hewes just sells 52 tickets.

a What is the probability of:

i Mr Raza winning the raffle

ii Mr Raza selling the winning ticket to someone else?

b What is the probability of either Mr or Mrs Raza selling the winning ticket to someone else?

c What is the probability of Mrs Hewes not selling a winning ticket?

d Which person has the greatest chance of either winning the raffle or selling the winning ticket? Explain your answer.

Give your answers as fractions in their simplest form.

This is a problem-solving question and so you must communicate your method clearly. Do not just write down probabilities without some explanation.	
a i $\dfrac{5}{400} = \dfrac{1}{80}$ **ii** $\dfrac{40}{400} = \dfrac{1}{10}$	Remember to cancel the fractions and make sure you read the information given in the question.
b $\dfrac{40}{400} + \dfrac{50}{400} = \dfrac{90}{400}$ $\qquad\qquad\quad = \dfrac{9}{40}$	Remember to show how you obtain the answer. It is usual to give it in its lowest terms.
c $1 - \dfrac{52}{400} = \dfrac{348}{400}$ $\qquad\qquad = \dfrac{87}{100}$	Remember that these are complementary outcomes.
d P(Mr Raza either winning the raffle or selling the winning ticket) $= \dfrac{45}{400}$ P(Mrs Raza either winning the raffle or selling the winning ticket) $= \dfrac{60}{400}$ P(Mrs Hewes either winning the raffle or selling the winning ticket) $= \dfrac{52}{400}$ Mrs Raza has the greatest chance, as $\dfrac{60}{400}$ is the largest fraction.	As you will need to compare fractions to solve the problem, there is no need to cancel the three probability fractions. Make sure you state your conclusion clearly and give a reason.

Susie is rehearsing for a driving test. This test is made up of two parts, a practical and a theory. She is told that the probability of passing only one of these two tests is 0.44 and the probability of passing the practical is 0.8.

a Draw a tree diagram to show this information.

b Set up an equation to calculate the probability of passing the theory test.

This is a problem-solving question. You must process the problem into a series of algebraic steps.	
a	Here, you are solving the problem by making a decision to put it into an algebraic context.
	Let x = P(passing the theory test) and so
	P(failing the theory test) = $1 - x$.
b P(only pass one test) = P(PF) + P(FP) $= 0.8(1 - x) + 0.2x$ So $0.8 - 0.8x + 0.2x = 0.44$ $0.8 - 0.6x = 0.44$ $\Rightarrow 0.6x = 0.8 - 0.44$ $x = \dfrac{0.8 - 0.44}{0.6}$ $= 0.6$	Use the tree diagram to work out P(only pass one test). Remember to add the two outcomes. This is the equation to solve. Rearrange the equation to calculate x.

Tree diagram (part a):

Practical	Theory	Outcome	Probability
Pass 0.8	x → Pass	PP	$0.8x$
	$1 - x$ → Fail	PF	$0.8(1 - x)$
Fail 0.2	x → Pass	FP	$0.2x$
	$1 - x$ → Fail	FF	$0.2(1 - x)$

 3 **a** A bag contains 10 red discs, 10 white discs and 10 blue discs.

Explain how to calculate P(3 red discs) if none are replaced.

b Harry has a bag that contains a quantity of red discs, white discs and blue discs.

He takes out ten discs at random, as a sample, and finds he has only blue discs and white discs. He says: 'This shows that there are no red discs in the bag.'

Is he correct? Give a reason to support your answer.

This question assesses your mathematical reasoning. You need to draw conclusions from the given information and demonstrate your mathematical understanding.	
a P(first disc is red) = $\dfrac{10}{30}$ $= \dfrac{1}{3}$ P(second disc is red) = $\dfrac{9}{29}$ P(third disc is red) = $\dfrac{8}{28}$ $= \dfrac{2}{7}$ So P(3 red discs) = $\dfrac{1}{3} \times \dfrac{9}{29} \times \dfrac{2}{7}$ $= \dfrac{6}{203}$ $= 0.0296$ (3 sf)	This is an example of conditional probability, as the discs are not replaced. You will need to show your working as part of the explanation. Your answer can be given as a fraction or a decimal. Here you are making a deduction to draw a conclusion from mathematical information.
b No, he is incorrect as there may be at least one red disc in the bag but he has not chosen them in the sample.	You need to say 'No' and give a valid, specific reason from given information.

6 Statistics

1. An opinion poll used a sample of 200 voters in one area. 112 said they would vote for Party A. There are a total of 50 000 voters in the area.

   a How many voters would you expect to vote for Party A if they all voted?

   b The poll is accurate to within 10%. Can Party A be confident of winning?

2. Sandila's school has 1260 students and there are 28 students in her class. A survey is carried out using students sampled from the whole school. Four boys and three girls in Sandila's class are chosen to take part in the survey.

   Estimate how many students in the whole school are in the sample.

3. You are asked to conduct a survey at a concert where the attendance is approximately 8000, which has roughly the same number of males and females.

   Explain how you could create a sample of the crowd.

4. An aunt tells you that the local train service is not as good as it used to be.

   a How could you find out if this is true?

   b Decide which data would be relevant to the enquiry and possible sources. How might you collect this data?

5. A coffee stain removed four numbers (in two columns) from the following frequency table of eggs laid by 20 hens one day.

Eggs	0	1	2		5
Frequency	2	3	4		1

   The mean number of eggs laid was 2.5.

   What could the missing four numbers be?

6. A hospital has to report the average waiting time for patients in the Accident and Emergency department. A survey was carried out to see how long patients waited before seeing a doctor.

   The table summarises the results for one shift. The times are rounded to the nearest minute.

Time (minutes)	0–10	11–20	21–30	31–40	41–50	51–60	61–70
Frequency	1	12	24	15	13	9	5

   a How many patients were seen by a doctor in this shift?

   b Estimate the mean waiting time per patient.

   c Which average would the hospital use for the average waiting time?

   d What percentage of patients did the doctors see within one hour?

7 Marion is writing an article on health for a magazine.

She asked a sample of people the question: 'When planning your diet, do you consider your health?'

The pie chart shows the results of her survey.

a Can you tell how many people there were in the sample? Give a reason for your answer.

b Write a brief summary of the article Marion could write using this data.

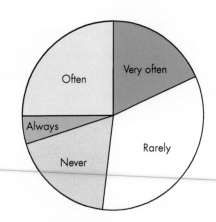

8 The frequency polygon shows the times that a number of people waited at the bus stop before their bus came one morning.

Dan says: 'Most people spent 5 minutes waiting.' Explain why this is incorrect.

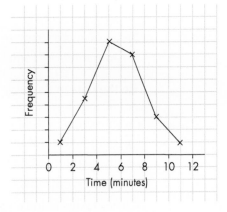

9 Andrew is asked to create a histogram. Explain to Andrew how he can work out the height of each bar on the frequency density scale.

10 Three dancers were hoping to be chosen to represent their school in a competition.

They had all been involved in previous competitions.

The table shows their scores in recent contests.

The teachers said they would be chosen by their best average score. Which average would each dancer prefer to be chosen by?

Kathy	8, 5, 6, 5, 7, 4, 5
Connie	8, 2, 7, 9, 2
Evie	8, 1, 8, 2, 3

11 Jasmin has just graduated from university. Two different companies are trying to convince Jasmin to join them as an employee. Company A has just sent her these figures.

Mean salary from firm A is £32 000 to the nearest thousand.

Mean salary from firm B is £28 000 to the nearest thousand.

Therefore it is best to have a job with firm A.

However, Jasmin has been talking to a graduate advisor from her university. The advisor showed Jasmin this table of salaries for the two companies.

Imagine you are the graduate advisor at the university. What advice would you give to Jasmin about the salaries of the two firms? Justify your recommendation to Jasmin.

Salaries from firm A	Salaries from firm B
£86 000	£45 000
£62 000	£36 000
£23 000	£26 000
£23 000	£26 000
£18 000	£26 000
£18 000	£26 000
£18 000	£22 000
£18 000	£22 000
£18 000	£22 000

12 Produce sets of grouped data with:

   a an estimated range of 26

   b an estimated median of 46

   c an estimated median of 22.5 and an estimated range of 62

   d an estimated mean of 36 (to one decimal place).

13 How would you make up a set of data with a median of 10 and an interquartile range of 7?

14 Connie planted some tomato plants and kept them in the kitchen, while her husband Harold planted some in the garden. After the summer, they compared their tomatoes.

	Connie	Harold
Mean diameter (cm)	1.9	4.3
Mean number of tomatoes per plant	23.2	12.3

Use the data in the table to explain who had the better crop of tomatoes.

15 **a** Which one of the following statements is more precise: 'My hypothesis is true,' or 'There is strong evidence to support my hypothesis.' Why?

   b Why might it not be the case that:

     i your hypothesis is true if you have found some evidence to support it

     ii you have failed if your hypothesis appears to be flawed.

16 Researchers at a market research company are trying to answer the question: 'Which newspaper is easiest to read?'

They carried out a newspaper survey of the number of letters in words in 500-word samples of a range of tabloid and broadsheet newspapers.

They calculated the mean number of letters in a word and the range.

Tabloid: mean 4.2 letters per word and range 9 letters per word

Broadsheet: mean 4.4 letters per word and range 14 letters per word

   a What do you think their survey has told them?

   b Do you think it provides them with sufficient evidence to answer their question?

   c What further question might you want to ask?

17 The table shows the total mass of fish caught by anglers in a fishing competition.

Mass (kg)	$0 < m \leqslant 5$	$5 < m \leqslant 10$	$10 < m \leqslant 15$	$15 < m \leqslant 20$	$20 < m \leqslant 25$
Frequency	4	15	10	8	3

Helen noticed that two frequencies have been swapped by accident and that this made a difference of 0.625 to the arithmetic mean.

Which two numbers are the wrong way round?

18 Marco is the head chef at Giorgio's restaurant. He regularly orders a range of cheeses to use in his recipes.

Here is a table showing the total masses of cheeses he has ordered over five years.

Cheese	Mass (kg)				
	2010	2011	2012	2013	2014
Wensleydale	26	34	33	35	38
Brie	17	13	16	21	22
Red Cheddar	13	17	21	22	26
White Cheshire	16	17	21	20	19
Red Leicester	13	12	16	22	23
Stilton	18	20	18	21	22
Edam	8	7	7	8	7

Marco wants to hand over responsibility for ordering cheeses to a new buying manager. The new manager has asked for some information.

Help Marco write a report about the cheeses ordered over these five years.

19 This table shows the number of cases of different types of wine bought in the UK over a five-year period.

Wine	Number of cases (millions)				
	2005	2006	2007	2008	2009
Chardonnay	6	8	8	8	9
Pinot grigio	2	3	4	5	5
Sauvignon blanc	3	4	5	5	6
Merlot	4	4	5	4	5
Shiraz	3	3	4	5	5
Cabernet sauvignon	4	5	4	5	5
Rioja	2	2	2	2	2

Write a report about the the sales of wines in the UK over these five years.

20 The lengths of time, in minutes, that it took for the Ambulance Service to get an ambulance to a patient and then to a hospital, are recorded.

The cumulative frequency diagram shows the data.

Calculate the estimated mean length of time it took the Ambulance Service to get to a patient and then to a hospital.

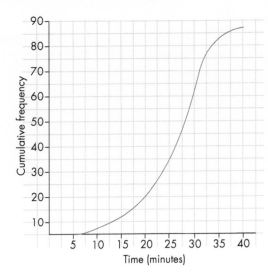

21 Johnny is given a cumulative frequency diagram showing the numbers of marks gained by students in a spelling test.

He is told that the top 15% are given the top grade.

How would you work out the marks needed to gain this top grade?

22 The histogram shows the science test scores for students in a school.

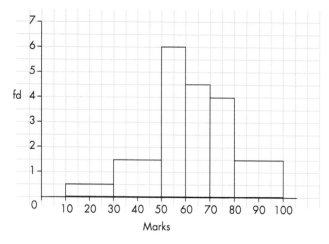

10% of the students gained a top grade. What score is needed for a top grade?

23 The distances that the members of a church travel to the church are shown in the histogram.

It is known that 22 members travel between 4 km and 6 km to church.

What is the probability of choosing a member at random who travels more than 4 km to church?

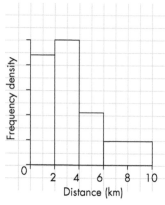

24 A dental practice has two dentists: Dr Ball and Dr Charlton. The box plots are created to illustrate the waiting times for their patients during November.

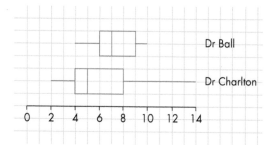

Gabriel is deciding which dentist to see.

Which one would you advise he see and why?

25 The box plots are for a school's end of year science tests.

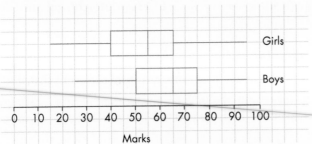

Write a brief summary on the test results for the next governors' meeting.

26 Joy is given the box plots for the daily sunshine in the seaside resorts of Scarborough and Blackpool for July, but no scale is shown.

She is told to write a report on the differences between the amounts of sunshine in both resorts.

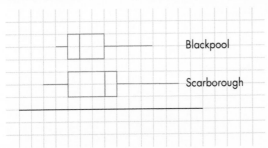

Write a report that could be justified from box plots with no scales shown.

27 Are these representations from different distributions? Explain your answer.

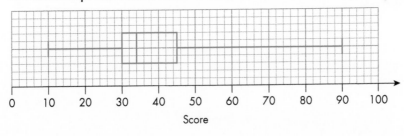

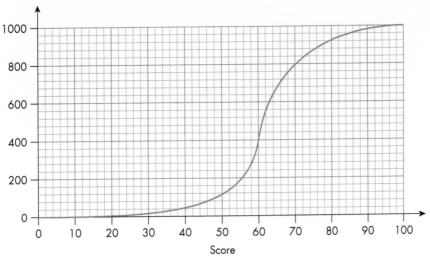

28 What would you expect to be the same/different about the two distributions representing the heights of students in years 7 to 13 and heights of students in years 12 and 13 only?

29 Look at the two distributions below.

a Explain how you know that the two diagrams are from different distributions.

b You are told they both represent age in years for woman at first marriage. One is for Europe and one is for Africa. Which do you think is which? Justify your decision.

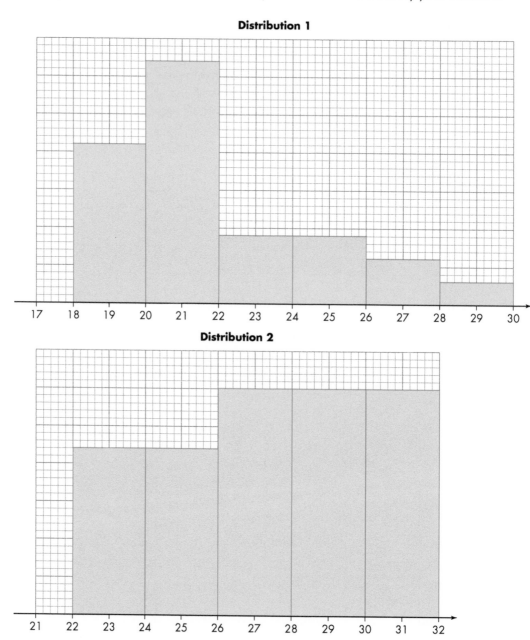

Distribution 1

Distribution 2

30 Data is described as skewed when the shape of its distribution has a long tail on one side or the other.

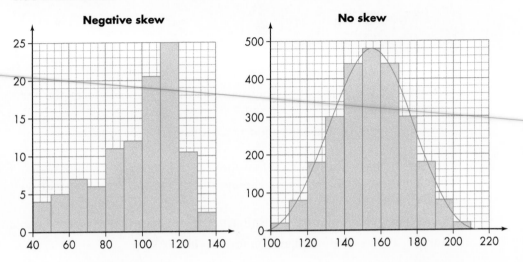

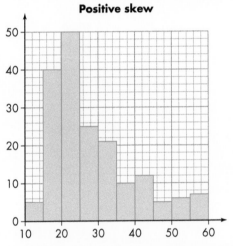

a Describe two contexts, one in which you might expect the variable to have a negative skew and the other in which it has positive skew.

b How can you tell from a box plot that the distribution has negative skew?

c Sketch a suitable box plot for each distribution.

31 The table shows data about air traffic between the UK and abroad from 1980 to 1990.

	1980	1982	1984	1986	1988	1990
Number of flights (thousands)	507	511	566	616	735	819
Number of passengers (millions)	43	44	51	52	71	77

a Draw a suitable diagram to illustrate the changes over this ten-year period.

b Research similar figures from the internet for the year 2000 and draw a suitable diagram to show the changes that have occurred between 1980 and 2000.

c Estimate the number of flights and number of passengers for the year 2010.

32 The table shows the time taken and distance travelled by a delivery van for 10 deliveries in one day.

Distance (km)	8	41	26	33	24	36	20	29	44	27
Time (min)	21	119	77	91	63	105	56	77	112	70

a Draw a suitable diagram.

b A delivery takes 30 minutes. How many kilometres would you expect the journey to have been?

c How long would you expect a journey of 30 kilometres to take?

33 Harry says he has a scatter graph that shows that in countries where people have more televisions they live longer. He says this means the more televisions you have the longer you live.

a Explain why Harry may have misinterpreted the information he has been given.

b Suggest a more likely explanation for the relationship.

⑦Hints and Tips

Question	Hint
4	How can 'good' be defined? Frequency of service, cost of journey, time taken, factors relating to comfort, access? How does the frequency of the train service vary throughout the day and week?
18, 19	Select the most appropriate statistical diagrams and measures to summarise the data given in the table.
29a	Consider the overall shape of the distributions and any statistical measure you can calculate or estimate from them.
33	Think about what other variables could be involved.

Worked exemplars

(PS) **1** The mean speed of each member of a cycling club over a long-distance race was recorded and a frequency polygon was drawn.

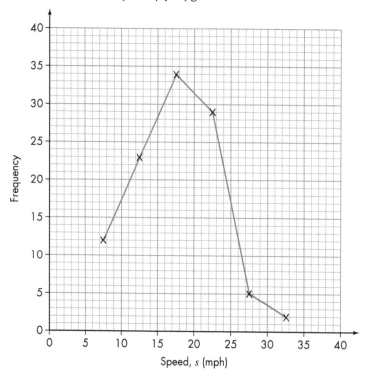

Speed, s (mph)

Use the frequency polygon to calculate an estimate for the mean speed.

This is a problem-solving question. You need to follow a series of processes to move the information from a graphical format into a tabular format.	
Create a grouped frequency table.	Remember that each plotted point is the mid-class value.

Speed, s (mph)	Frequency, f	Midpoint, m	f × m
5 < s ≤ 10	12	7.5	90
10 < s ≤ 15	23	12.5	287.5
15 < s ≤ 20	34	17.5	595
20 < s ≤ 25	24	22.5	540
25 < s ≤ 30	5	27.5	137.5
30 < s ≤ 35	2	32.5	65
	100		1715

Estimate for the mean speed = 1715 ÷ 100	Use the information from your table to calculate an estimate for the mean speed to a suitable degree of accuracy.
= 17.15	
= 17.2 mph (1 dp)	

2 Simon makes men's and women's shirts. He needs to know the range of collar sizes so he measures 100 men's necks. The results are shown in the table.

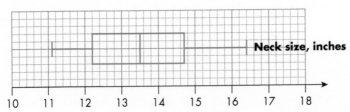

Neck size, n (inches)	Frequency
$12 < n \leqslant 13$	5
$13 < n \leqslant 14$	16
$14 < n \leqslant 15$	28
$15 < n \leqslant 16$	37
$16 < n \leqslant 17$	10
$17 < n \leqslant 18$	4

a Draw a cumulative frequency graph to show this information.

b Use the graph to work out:

   i the median   **ii** the interquartile range.

c The box plot shows the neck sizes of 100 women.

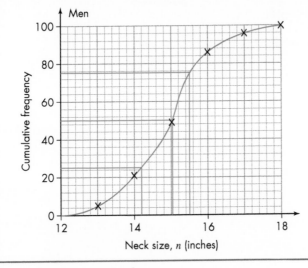

Compare the distribution of neck sizes for men and women.

> This is a mathematical reasoning question. You need to demonstrate your use of mathematical skills and knowledge in your answer.

a The cumulative frequencies are: 5, 21, 49, 86, 96, 100.

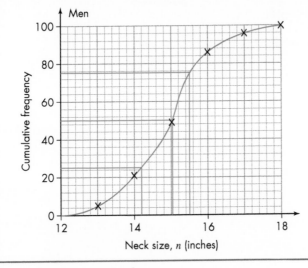

Remember to plot the cumulative frequencies at the end of each class interval.

You can draw either a cumulative frequency polygon or a cumulative frequency curve.

b i Median = 15 inches

  ii Lower quartile = 14.2 inches

    Upper quartile = 15.5 inches

    So IQR = 15.5 – 14.2 = 1.3 inches

You need to read these values by drawing lines from the relevant places on the cumulative frequency axis: 50, 25 and 75.

c For women:

  median = 13.5 inches

  IQR = 14.7 – 12.2 = 2.5 inches

  Comparing the medians shows that, on average, men have larger neck sizes than women.

  Comparing the IQRs show that the neck sizes of the women are more spread out.

You must carefully compare the medians and the IQRs and make a written conclusion.

Here you are making a deduction to draw a conclusion from mathematical information.

William Collins' dream of knowledge for all began with the publication of his first book in 1819. A self-educated mill worker, he not only enriched millions of lives, but also founded a flourishing publishing house. Today, staying true to this spirit, Collins books are packed with inspiration, innovation and practical expertise. They place you at the centre of a world of possibility and give you exactly what you need to explore it.

Collins. Freedom to teach

Published by Collins
An imprint of HarperCollins*Publishers*
News Building
1 London Bridge Street
London SE1 9GF

Browse the complete Collins catalogue at
www.collins.co.uk

© HarperCollins*Publishers* Limited 2015

10 9 8 7 6 5 4 3 2

ISBN 978-0-00-811385-8

A catalogue record for this book is available from the British Library

The author Sandra Wharton asserts her moral rights to be identified as the author of this work.

Commissioned by Lucy Rowland and Katie Sergeant
Project managed by Elektra Media Ltd and Hart McLeod Ltd
Contributions from Jo-Anne Lees and Brian Speed
Copyedited by Jim Newall
Proofread by Joan Miller
Answers checked by Jim Newall
Edited by Jennifer Yong
Typeset by Jouve India Private Limited
Illustrations by Ann Paganuzzi
Designed by Ken Vail Graphic Design
Cover design by We are Laura
Production by Rachel Weaver

Acknowledgements
The publishers gratefully acknowledge the permissions granted to reproduce copyright material in this book. Every effort has been made to contact the holders of copyright material, but if any have been inadvertently overlooked, the publisher will be pleased to make the necessary arrangements at the first opportunity.

The publishers would like to thank the following for permission to reproduce photographs in these pages:

Cover (bottom) Procy/Shutterstock, cover (top) joingate/Shutterstock.